The Palgrave Macmillan Animal Ethics Series

Series Editors
Andrew Linzey, Oxford Centre for Animal Ethics, Oxford, UK
Clair Linzey, Oxford Centre for Animal Ethics, Oxford, UK

In recent years, there has been a growing interest in the ethics of our treatment of animals. Philosophers have led the way, and now a range of other scholars have followed from historians to social scientists. From being a marginal issue, animals have become an emerging issue in ethics and in multidisciplinary inquiry. This series will explore the challenges that Animal Ethics poses, both conceptually and practically, to traditional understandings of human-animal relations. Specifically, the Series will:

- provide a range of key introductory and advanced texts that map out ethical positions on animals;
- publish pioneering work written by new, as well as accomplished, scholars;
- produce texts from a variety of disciplines that are multidisciplinary in character or have multidisciplinary relevance.

For further information or to submit a proposal for consideration, please contact Amy Invernizzi, amy.invernizzi@palgrave-usa.com.

Daniela Rizzo

Animal Pneumatology

A Systematic Theological Inquiry into the Spirit's Work in Non-Human Creatures

palgrave
macmillan

Daniela Rizzo
Alphacrucis University College
Parramatta, NSW, Australia

ISSN 2634-6672 ISSN 2634-6680 (electronic)
The Palgrave Macmillan Animal Ethics Series
ISBN 978-3-032-18722-2 ISBN 978-3-032-18723-9 (eBook)
https://doi.org/10.1007/978-3-032-18723-9

Cover credit: Daniela Rizzo

This Palgrave Macmillan imprint is published by the registered company Springer Nature Switzerland AG
The registered company address is: Gewerbestrasse 11, 6330 Cham, Switzerland

For my mother, Marisa, whose life breathes the gentleness and strength of the Holy Spirit, and whose faith has been a guiding presence in mine.

Series Editors' Preface

This is a new book series for a new field of inquiry: Animal Ethics.

In recent years, there has been a growing interest in the ethics of our treatment of animals. Philosophers have led the way, and now a range of other scholars have followed from historians to social scientists. From being a marginal issue, animals have become an emerging issue in ethics and in multidisciplinary inquiry.

In addition, a rethink of the status of animals has been fuelled by a range of scientific investigations which have revealed the complexity of animal sentiency, cognition and awareness. The ethical implications of this new knowledge have yet to be properly evaluated, but it is becoming clear that the old view that animals are mere things, tools, machines or commodities cannot be sustained ethically.

But it is not only philosophy and science that are putting animals on the agenda. Increasingly, in Europe and the United States, animals are becoming a political issue as political parties vie for the "green" and "animal" vote. In turn, political scientists are beginning to look again at the history of political thought in relation to animals, and historians are beginning to revisit the political history of animal protection.

As animals grow as an issue of importance, so there have been more collaborative academic ventures leading to conference volumes, special journal issues, indeed new academic animal journals as well. Moreover, we have witnessed the growth of academic courses, as well as university posts, in Animal Ethics, Animal Welfare, Animal Rights, Animal Law, Animals and Philosophy, Human-Animal Studies, Critical Animal Studies, Animals and Society, Animals in Literature, Animals and Religion—tangible signs that a new academic discipline is emerging.

"Animal Ethics" is the new term for the academic exploration of the moral status of the non-human—an exploration that explicitly involves a focus on what we owe animals morally, and which also helps us to understand the influences—social, legal, cultural, religious and political—that legitimate animal abuse. This series explores the challenges that Animal Ethics poses, both conceptually and practically, to traditional understandings of human-animal relations.

The series is needed for three reasons: (i) to provide the texts that will service the new university courses on animals; (ii) to support the increasing number of students studying and academics researching in animal related fields, and (iii) because there is currently no book series that is a focus for multidisciplinary research in the field.

Specifically, the series will

- provide a range of key introductory and advanced texts that map out ethical positions on animals;
- publish pioneering work written by new, as well as accomplished, scholars, and
- produce texts from a variety of disciplines that are multidisciplinary in character or have multidisciplinary relevance.

The new Palgrave Macmillan Series on Animal Ethics is the result of a unique partnership between Palgrave Macmillan and the Ferrater Mora Oxford Centre for Animal Ethics. The series is an integral part of the mission of the Centre to put animals on the intellectual agenda by facilitating academic research and publication. The series is also a natural complement to one of the Centre's other major projects, the *Journal of Animal Ethics*. The Centre is an independent "think tank" for the advancement of progressive thought about animals, and is the first Centre of its kind in the world. It aims to demonstrate rigorous intellectual enquiry and the highest standards of scholarship. It strives to be a world-class centre of academic excellence in its field.

We invite academics to visit the Centre's website www.oxfordanimalethics.com and to contact us with new book proposals for the series.

Oxford, UK
Oxford, UK

Andrew Linzey
Clair Linzey
General Editors

Foreword

Karl Barth, arguably the greatest theologian of the twentieth century, produced more than 12 huge volumes of his magnum opus, the *Church Dogmatics*. Yet, his work was never completed. Towards the end of his life, he dreamt of the possibility of another approach—a "theology of the Holy Spirit"—"a theology which now I can only envisage from afar, as Moses once looked on the promised land."

"He was thinking," writes his biographer, Eberhard Busch, "of a theology which, unlike his own, was not written from the dominant perspective of Christology, but from that of Pneumatology." (*Karl Barth: His Life from Letters and Autobiographical Texts*, SCM Press, 1976, p. 494.)

Now Daniela Rizzo takes up this challenge in relation to Christian thinking not only about humans, but also about the world of fellow creatures, a field which has been largely dominated by Christological approaches. The quality of this work on animal Pneumatology is stunning. It is outstanding, original, and ground-breaking. I have worked in the field of animal theology for more than half a century, and I have never read a more brilliant engagement with this topic.

Her constructive work on the spiritual life of animals raises the bar of theological reflection on the lives of animals, and her call (albeit rather modest) for Christian communities to take seriously the work of the Spirit in relation to individual creatures (as well as creation as a whole) is unassailable.

I have spent hours pondering this work and am delighted to have lived long enough to read such a first-class development in the field in which I have worked for so long. This is a book that will galvanise further scholarship and reflection worldwide.

Oxford, UK Andrew Linzey

The Revd Professor Andrew Linzey, Ph.D., D.D., Hon.D.D., is Director of the Oxford Centre for Animal Ethics, and has been a member of the Faculty of Theology in the University of Oxford for 28 years. He has written or edited over 40 books, including *Animal Theology* (SCM Press/University of Illinois Press, 1994), *Why Animal Suffering Matters* (Oxford University Press, 2009), and *Animal Theologians* (Oxford University Press, 2023).

Preface

It was not in the soaring hymns of a cathedral or the pages of a theological tome that the Spirit found me, but in the hush of a house heavy with mourning. My father had died. I had been praying (really, pleading) for some trace of God to appear amid the grief, a whisper in the dark. What came instead was the soft weight of my cat, Pacino (pictured on the cover of this book). He climbed onto me as I sat defeated on the sofa, pressed his head against my chest, and stayed. There were no words, no visions, no miracles. Only the slow, steady rhythm of his breathing synced with mine, yet it felt like the Spirit's tangible comfort. Not mere companionship, but communion. As if the Spirit who hovered over the deep had folded himself into fur and bone.

That quiet moment could easily have slipped into the margins of memory, but it became a seed. Could the Spirit move not only through prophets and preachers, but through purrs? Could the breath of God fill not just human lungs, but the creatures we overlook in our search for grander signs of the divine?

I couldn't shake these questions. Instead, they became the heartbeat of my theological work: a pneumatology of animals. It was a way of seeing creatures not as background to the human story but as Spirit-breathed participants in God's living world. Scripture, read with widened eyes, makes space for this. Tradition, when freed from its anthropocentric bindings, echoes it. I have come to believe there is a Spirit-thread running through all flesh, and animals, too, are drawn along by the Spirit's breath.

This conviction has only grown sharper in light of the ecological crisis of our time, and it is sharpened again by my own context as an Australian Pentecostal theologian. Here, the beauty and fragility of biodiversity, the urgency of ecological crisis, and the responsibilities of stewardship press upon theological reflection. These realities call us to recognise that creation care is not an optional concern, but one that lies at the very heart of Christian life. Pentecostalism has championed the experiential work of the Spirit in human life but seldom extended that imagination to the non-human creature. My hope in these pages is to bring pneumatology, animal theology, and Pentecostal theology into closer conversation, and to do so in a way that honours both the Spirit's sustaining presence and the creatures caught up in it.

What follows, then, is not only an academic project but a deeply personal one. I hope it will invite readers to see animals differently: not as decorative to human existence, but as creatures alive in the Spirit's sustaining presence. My prayer is that these pages might deepen our understanding of the Spirit's work in creation and stir us towards more faithful, more compassionate ways of living with all of God's creatures. A personal witness to a God whose Spirit moves where it will, sometimes in a whisper, sometimes in a breath, and sometimes in the quiet presence of an animal who comforts.

Parramatta, Australia Daniela Rizzo

Acknowledgements When I first set out to write this book, the task seemed nearly impossible: how does one name the movement of the Holy Spirit in the lives of animals and translate unarticulated breath into words? Though born of doctoral research, this work has been shaped as much by conversation, critique, and companionship as by solitary study. I am indebted to all who, in their wisdom and generosity, made these pages possible.

I owe a special debt of gratitude to my primary supervisor, Grant Buchanan. He taught me to dream theology beneath star-lit skies, to trust imagination as a theological method, and to allow the Pentecostal experience to open new vistas of thought. His confidence in the project and his encouragement to pursue its convictions gave this work its imaginative lift. My secondary supervisor, Ernst Conradie, brought the gift of rigorous discipline and critical clarity. His uncompromising standards refined the argument, secured its theological coherence, and pressed me beyond the boundaries of my comfort.

I am thankful to colleagues who, in different seasons, spoke into this research. DJ Konz, U-Wen Low, and John Griffiths offered counsel and critique that shaped the contours of my argument and sustained its hope. Jacqueline Grey, Kerrie Stevens, Danielle Terceiro, Anita Davis, Kay Doyle, Danielle Hopkins, and Mark Jacobson provided practical support in varying forms, such as mentorship, provision, friendship, resources, and encouragement that carried me through demanding stretches of work. I am especially grateful to Ben Myers, who believed in the project from the outset, investing time, editorial guidance, and research opportunities that profoundly marked my development as a scholar as I completed this work.

To members of the Pentecostal scholarly community, I owe a further debt. Frank Macchia and Amos Yong read my original thesis with characteristic generosity, offering feedback that was both exacting and life-giving. To Andrew Linzey, whose lifelong advocacy for animals has embodied the vocation of Christian scholarship, I extend special thanks. Your sacrificial witness and courage have been a beacon, and it has been an honour to write this first pneumatological treatment of animal theology in the light of your example. I thank both you and Clair for welcoming me with open arms into the community that is the Oxford Centre for Animal Ethics.

My family has been the sure foundation beneath this work. To my husband, Davide: thank you for steadying me in moments of uncertainty and for walking patiently beside me to the final line—your love carried me through. To my mother, Marisa, my siblings, Michael and Gabriella, and my brother-in-law, Uilao: you have believed in this calling from the very beginning. Your prayers, practical care, wise words, and unwavering encouragement sustained me in ways beyond words. My dear friends, Les and Sue Gardiner, thank you both for your comfort, good humour, and listening ears through the highs and lows of this season. I also take this opportunity to remember my father, Vince Rizzo, a church pastor, lover of Scripture, and of rottweilers, whose presence is no longer with us: your determination to finish what was begun lingers still, and I write in the light of that legacy.

Finally, to Pacino—my constant feline companion. From the first sentence to these acknowledgements, you have been faithfully at my side. Thank you for comforting me in my season of mourning and for inspiring the life-giving words in this book. In your quiet presence, I have glimpsed the eschatological promise that runs through these pages: that in the fullness of God's redemption, every creaturely voice shall be heard and reconciled.

To all named here, and to those unnamed yet no less dear, my deepest thanks. This book is the fruit of communal labour; may it be worthy of the life, the breath, and the flourishing of all God's creatures.

Competing Interests The author has no competing interests to declare that are relevant to the content of this manuscript.

Praise for *Animal Pneumatology*

"If human flesh at Pentecost received power to bear witness to the ends of the Earth, Daniela Rizzo's attentiveness to the full scriptural testimony to animals shows that their witness via the primordial infusion of divine breath has already resonated cosmically even in and through the appearance of adamic failure. *Animal Pneumatology* prophetically announces how the sanctifying work of Yahweh unfolded on the creational week not just anticipates human redemption but also heralds the incarnational and spirited worship of all God's living creatures around the eternal throne. Carefully read and then even more meticulously reread this book: herein the many tongues of the pneumatological imagination are unleashed for our age of the Anthropocene!"

—Amos Yong, *Professor of Theology and Mission, Fuller Theological Seminary, USA*

"Rizzo dismantles anthropocentric thinking by arguing persuasively that animals are active participants alongside humanity in the divine liberation of creation. Her book raised provocative and significant theological issues for me that I should have long ago taken more seriously. I have become deeply grateful for her voice."

—Frank D. Macchia, *Professor of Systematic Theology, Vanguard University; Associate Director of the Centre for Pentecostal and Charismatic Studies, Bangor University, Wales, UK*

"Most theologians agree that God's creation includes 'all flesh,' though most often in general but regrettably seldom in particular. This book, in contrast, offers an unprecedented counterbalance by focussing our attention to the sentient animals as theological subjects. In an exceptional way, the author deepens existing animal theologies with the rich diversity of pneumatology, inventively acknowledging the animals as co-creatures of the Holy Spirit in the Triune's history of salvation. I am sure the reader will face animals in a new caring way through this thought-provoking work."

—Sigurd Bergmann, *Professor emeritus in Religious Studies, Trondheim/Uppsala/Munich/Lund; Secretary of the European Forum for the Study of Religion and the Environment*

"In this book Rizzo breaks exciting new ground. Here God and Earth, divinity and animality, revelation and carnal existence are one. Challenging the conventional Christian dualism between Creator and creation, Rizzo envisions the natural world as enlivened by the Spirit in its solidarity with all the vulnerable and suffering denizens of Mother Earth. A must for readers at the intersection between new animal studies and the importance of the Holy Spirit in Christian thought."

—Mark Wallace, *James Hormel Professor of Social Justice, Department of Religion, Swarthmore College, USA*

"The animal question is a key challenge of our time, with profound implications for Christian doctrine and practice. Rizzo's book is an important contribution to the growing field of theological engagement with animals, addressing a significant lacuna in previous work in the field through a careful and well-informed exploration of pneumatology."

—David L. Clough, *Chair in Theology and Applied Sciences, University of Aberdeen, UK*

"What a beautiful, enlightening book! Laced with biblical, theological, ethical, and scientific insights, its exploration into the relationship between the Spirit and animals is positively eye-opening. Knowledgeable about today's debates, Rizzo imbues the discussion with unusual warmth and keenness due to her own participation in the Pentecostal tradition.

Among the welcome sea of books on ecological theology, this book stands out as a genuinely original must-read. There is simply nothing else like it."

—Elizabeth A. Johnson, *Distinguished Professor of Theology Emerita, Fordham University, USA*

Contents

About the Author

Daniela Rizzo is a Lecturer in Systematic Theology at Alphacrucis University College (Australia), whose research explores the intersection of systematic pneumatology and animal studies. She is a Fellow of the Oxford Centre for Animal Ethics, with academic memberships including the Australasian Animal Studies Association, A Rocha Australia, and the Pentecostal World Fellowship Creation Care Task Force. In 2025, Daniela received both the Alphacrucis President's Doctoral Research Medal and the inaugural Ross Winchester Award for Pentecostal Studies in recognition of her innovative contributions to Pentecostal theological research.

Residing along Australia's iconic Great Ocean Road, Daniela draws inspiration from the landscape she inhabits—a place alive with penguins, whales, wallabies, koalas, and kangaroos. This vibrant ecological context shapes her theological imagination, grounding her work in the lived reality of creaturely life and the Spirit's presence within the more-than-human world.

CHAPTER 1

Where the Spirit Meets the Creature

To speak of the Spirit and the animal together is to enter a conversation that theology has only just begun. This book approaches that conversation through the lens of pneumatology, offering a doctrinally expansive and theologically original account of non-human life in relation to the Spirit. Weaving together historical and contemporary voices, it traces how animals have been imagined (or overlooked) within theology, while pressing into the unanswered questions of the Spirit's presence in, with, and through creaturely life. At the heart of this work is a question: what shape might a pneumatology of animals take, and how might such a vision reframe our understanding of creation, agency, and the presence of the Spirit?

While the growing field of animal theology has brought attention to theological considerations and ethical concerns regarding animals, it has only made limited reference to pneumatology—the doctrine of the Holy Spirit. This is a surprising oversight, as both classical and contemporary Christian theologies, particularly from the twentieth and twenty-first centuries, have emphasised the centrality of the Spirit in the acts of creation, as well as in sustaining and renewing the world. Engaging pneumatology in animal theology is not merely an exercise in filling a gap; rather, it is a crucial theological endeavour that opens up a richer, Spirit-centred understanding of animals within God's creation. Without such a perspective, key dimensions of theological reflection on animals remain

D. Rizzo, *Animal Pneumatology*, The Palgrave Macmillan Animal Ethics Series, https://doi.org/10.1007/978-3-032-18723-9_1

underdeveloped, including their participation in the life-giving work of the Spirit, which extends beyond human-centred concerns. A focused theological treatment of animals from a pneumatological perspective is thus overdue.

A pneumatological approach to animal theology is vital because it allows for a vision of animals not merely as created beings but as active participants in the Spirit's ongoing work in the world. By focusing on the Spirit's dynamic and relational presence, this approach can transform how we understand animals—not as passive recipients of care but as co-inhabitants with intrinsic value, animated and sustained by the same Spirit that gives life to all of creation. This perspective is particularly relevant within the Pentecostal tradition, which emphasises the Spirit's immediacy and active presence. Pentecostal pneumatology, with its focus on the Spirit's transformative power, presents a unique theological framework through which to understand the relationship between the Spirit and the non-human world, offering a fresh and impactful contribution to both animal theology and ecotheology.[1]

[1] A growing theological disconnect between Pentecostalism's early revolutionary spirit and its contemporary disengagement from environmental issues has been a focal point of critique within Pentecostal scholarship. Shane Clifton observes that "environmental concern has rarely, if ever, been an important element of the Pentecostal message." Shane Clifton, "Preaching the 'Full Gospel' in the Context of Global Environmental Crisis," in *The Spirit Renews the Face of the Earth: Pentecostal Forays in Science and Theology of Creation*, ed. Amos Yong (Eugene, OR: Pickwick Publications, 2009), 117. This critique is not new; as early as 1978, Walter Hollenweger—often regarded as the Dean of Pentecostalism—raised concerns about the movement's lack of attention to ecological issues. Hollenweger's scholarship was pivotal in articulating the significance of Pentecostal and Charismatic movements within global Christianity, reshaping ecumenical dialogue, and renewing the theological value of oral tradition. A. J. Swoboda, "Greening Hollenweger: Retrieving Ecological Dimensions from the Dean of Pentecostal Studies," in *Pentecostal Missiology and Environmental Degradation*, ed. Eugene Baron and Amos Yong (Carlisle, UK: Langham Global, Forthcoming).

Hollenweger was particularly critical of how early Pentecostalism and evangelicalism, which had once been revolutionary in their advocacy for social and political change—such as antislavery efforts, female leadership, and alternative economic practices—had gradually distanced themselves from these transformative roots.. See Cheryl J Sanders, "Pentecostal Ethics and the Prosperity Gospel: Is There a Prophet in the House?," in *Afro-Pentecostalism: Black Pentecostal and Charismatic Christianity in History and Culture*., ed. Estrelda. Y Alexander and Amos Yong (New York: New York University Press, 2011), 142–43. Hollenweger highlighted the irony that a movement once characterised by radical social engagement had, over time, grown more sectarian and disengaged from worldly affairs. In a later interview, Hollenweger identified two key challenges for Pentecostalism's future. First, he expressed concern over its increasing sectarianism and

Moreover, a pneumatology of animals provides a necessary corrective to the predominantly Christological and anthropocentric frameworks that have historically overshadowed the Spirit's role in creation. The Spirit's work in sustaining and animating all life (not just human life) calls for a broadened theological imagination, one that affirms the Spirit's presence in all creatures and redefines their place within the divine economy. In this way, pneumatology can expand theological discourse by emphasising the Spirit's inclusivity and relationality, inviting a more comprehensive view of animals as integral participants in creation. The urgency of developing this pneumatological approach is underscored by the contemporary ecological crisis, which threatens both the natural world and its non-human inhabitants. A pneumatological lens not only highlights the Spirit's sustaining and renewing role in creation but also frames the care for animals as a spiritual imperative. In doing so, it challenges the Christian to respond ethically and theologically to ecological destruction, situating animal care within a broader framework of creation care that reflects a commitment to the Spirit's life-giving work.

withdrawal from public life. Second, he questioned how Pentecostalism, and Christianity more broadly, would respond to the ecological crisis, asking, "What do we do with the ecological threat to the world? What do we do with the threat of hunger and the plight of refugees?" Clifton, "Preaching the 'Full Gospel' in the Context of Global Environmental Crisis," 118. Hollenweger's reflections drew a clear connection between ecological and humanitarian concerns, emphasising that the poor are disproportionately affected by environmental degradation. He argued that Pentecostalism needed a pneumatology that went beyond ecclesiology—a pneumatology that recognised the Spirit as the "Spirit of Life," who values, connects, and sustains all creation. See W. J. Hollenweger, "Creator Spiritus: The Challenge of Pentecostal Experience to Pentecostal Theology," *Theology* 81, no. 679 (1978): 32–40, https://doi.org/10.1177/0040571X7808100106. For Hollenweger, addressing the ecological crisis was inseparable from renewing a commitment to social justice. His broader critique of Pentecostalism's departure from its early commitment to social and political transformation suggests that ecological concerns must be understood as part of the movement's mandate for justice. In the broader discussion of Evangelical engagement with environmental and animal welfare issues, the historical contributions of faith communities have often been overlooked. While prevailing narratives tend to emphasise secular or Enlightenment influences, it is crucial to acknowledge the significant role Christian, and particularly Evangelical, theology has played in shaping these movements. Contrary to the perception that Christianity has been indifferent or antagonistic towards environmental and animal welfare, Evangelicals were among the early advocates for the protection of animals and for moral responsibility toward creation. Their theological commitment to stewardship and care for the vulnerable, both human and non-human, has been a foundational aspect of what is now considered humanitarian progress. See my chapter, Rizzo, D. "Pentecostal and Evangelical Ecotheologies", in Eaton, M. & Harvie, T., *The Routledge Handbook of Ecotheology. Forthcoming.*

In these pages, I set out to imagine a pneumatology of animals within the breadth of a Trinitarian vision, drawing on Pentecostal and ecumenical voices. Such an approach does more than fill a gap in animal theology; it invites a richer conversation about creation care and ecological responsibility, where the Spirit is seen as the breath that animates and sustains all life.

The Journey Ahead

This book sets out to chart new terrain in theology by exploring the pneumatological dimensions of non-human animal life and agency. Rooted in the Christian tradition and animated by biblical visions of the Spirit as giver of life, it asks what it might mean to recognise the Spirit as present and active in the lives of animals. Here, animals are not framed as passive recipients of God's creative breath, but as participants in the Spirit's ongoing work—creatures with spiritual significance in their own right.

Rather than merely filling a gap in the field, this work offers an innovative reframing of pneumatology. By bringing Spirit-baptism and glossolalia into conversation with the lives of animals, it extends the horizon of theological imagination beyond human experience. Such a move disrupts entrenched anthropocentric frameworks and widens the scope of pneumatology to encompass the whole creation. The implications are as practical as they are theological: if the Spirit is active in non-human lives, then Christian ethics must reckon with animals not as peripheral but as participants in God's redemptive purposes.

In constructing this vision, I engage a wide constellation of voices. Animal theologians such as Andrew Linzey and David Clough have long pressed for ethical seriousness regarding animals, while Jürgen Moltmann and Denis Edwards provide pneumatological insights that tether the Spirit's work to creation and renewal. Pentecostal scholars like Frank Macchia and A. J. Swoboda contribute a distinctive lens, enriching this exploration with the vitality and immediacy of Pentecostal spirituality.

The implications of such a pneumatology are profound. It calls for nothing less than a reorientation of theology and practice: a vision of creation care in which the Spirit's sustaining, renewing, and healing presence is discerned not only in human life but in the life of every creature. To see animals as caught up in this Spirit-filled economy is to deepen theology's account of creation and to press Christian communities towards more faithful, more holistic, and more compassionate practices of care.

The shape of this vision emerges through a systematic pneumatology of animals, traced across chapters that consider Creation and Covenant, Predation and Sin, Soteriology, Eschatology, and Glossolalia. Each chapter addresses a distinct aspect of pneumatology as it relates to the lives of non-human creatures, with the aim of offering a comprehensive theological framework that accounts for the Spirit's presence and activity in all of creation.

Chapter 2, on Creation and Covenant, explores the pneumatological foundations of animal life, focusing on the biblical concepts of *ruach* (Spirit or breath) and *nephesh* (soul or life force) as they apply to animals. Traditionally, these terms have been interpreted within human-centred frameworks, but here, they are presented as inclusive principles that animate all living beings, suggesting that animals are recipients of divine breath and integral participants in a Spirit-infused creation. By establishing *ruach* and *nephesh* as foundational to both human and non-human life, the chapter provides a theological basis for viewing animals as relational beings with potential for spiritual engagement.

From this groundwork, the discussion moves into examining animal sentience (their capacity to experience emotions and subjective states) as an extension of their Spirit-filled existence. If animals share the same divine breath that animates humans, it follows that they might possess affective depth and relational capacities that align with theological understandings of spirituality. This shift from life force to sentience expands theological perspectives on animals, recognising their potential to experience awe, mourning, and relational connection, which can be understood as elements of spirituality. The chapter then moves further into the realms of animal communication and ontology, suggesting that animals, through their unique expressions and behaviours, engage in forms of relationality that reflect an intrinsic connection to the Spirit. Scientific research, such as observations of animal mourning rituals or intricate social communications, supports the idea that animals possess modes of interaction that are not simply biological but indicative of a relational life. These communicative acts serve as expressions of animals' inner lives, possibly resonant with divine relationality, challenging anthropocentric notions of spiritual interaction.

This progression from *ruach* and *nephesh* to sentience, communication, and, ultimately, ontology reframes animals as creatures with intrinsic worth and a relational existence that aligns them with God's covenantal relationship with all creation. Building on this expanded pneumatology that includes animals as integral participants in creation, the chapter transitions to explore predation and natural suffering. It considers how the Holy Spirit engages with the realities of animal suffering and mortality, challenging anthropocentric views by framing natural evil as an experience shared across all of creation.

Chapter 3, on Predation and Sin, considers the role of the Holy Spirit within the realities of suffering, mortality, and moral complexity in the natural world, specifically addressing how non-human animals experience these phenomena. It begins by defining natural evil as harm that occurs independently of human agency, such as predation and environmental suffering. This categorisation serves to challenge the anthropocentric view that confines sin, suffering, and moral agency to humans alone, proposing instead that non-human creatures also partake in creation's intricate experiences of pain and survival. The chapter unfolds through a theological exploration of predation, questioning traditional interpretations that link suffering to a fallen creation. It contrasts two primary theological perspectives: one that views predation as an inherent aspect of an unfallen creation, and another that attributes it to a cosmic Fall. The chapter argues that predation may reflect God's intentional design, viewing evolutionary processes—including death and suffering—as expressions of divine love and complexity. This perspective challenges idealised notions of a prelapsarian, harmonious creation. Conversely, some theologians argue that predation and suffering are outcomes of a corrupted, fallen creation. This view posits that such violence is incompatible with God's original intent and that redemption through Christ aims to restore creation's intended harmony. The theological tension between these perspectives sets the stage for deeper questions about how to interpret the presence of natural evil within divine creation. The discussion then considers animal moral agency, reflecting on whether animals can be morally accountable for their actions or if their behaviours reflect a natural, non-moral state. I explore the possibility of proto-morality in animal behaviour, which, while not identical to human moral agency, reflects social behaviours and relational values intrinsic to animals' existence.

A pneumatological approach forms the core of the chapter, proposing that the Holy Spirit's presence permeates the natural order, including predation and suffering. The Spirit is portrayed as the sustainer of life, actively guiding and transforming creation through its complex and often painful realities. Rather than eliminating predation, the Spirit works within these processes, fostering ecological balance and promoting biodiversity. I propose that the Spirit's role encompasses both the vitality and suffering of non-human life, framing predation not solely as a manifestation of natural evil but as an integral aspect of the Spirit's creative and sustaining work. The chapter invites readers to consider a pneumatology that recognises the Spirit's engagement with creation's challenging aspects. It posits that the Spirit is dynamically present amid the struggles of animal existence, not to eradicate suffering but to guide creation towards ultimate reconciliation. Through this perspective, the chapter opens a dialogue on the Spirit's redemptive presence within both the beauty and brutality of the natural world, acknowledging the Spirit's work even within the violence and suffering of evolutionary processes.

By framing the Spirit as actively present within creation's hardships, chapter 4 turns from Christ-centred discussions to a distinctively pneumatological perspective on animal salvation. While honouring the foundational contributions of scholars like Andrew Linzey and David Clough—who have emphasised the cosmic reach of Christ's incarnation—chapter 4 on Soteriology builds on their work by focusing on the Spirit's particular role in the divine embrace of all creation, especially non-human creatures. Through this pneumatological lens, the Spirit's presence is seen as a life-giving, sanctifying force that extends salvation to "all flesh," an inclusivity that incorporates animals as active members within the divine economy of redemption. In this chapter, I examine theological tensions regarding animal suffering and question the traditional boundary that reserves salvation for humanity alone. I argue that animals are participants in the Spirit's redemptive mission, embodying the Creator's compassion and shared kinship in the divine plan.

I then build on the idea of "deep incarnation," where Christ's identification with "flesh" encompasses all creation, to include a robust pneumatological dimension. This view moves beyond an anthropocentric interpretation to a broader vision of the Word's embodiment that includes animals within a divine kinship. Introducing the Spirit's work into this framework, the chapter proposes that the Spirit's "baptism of all flesh" fosters a transformative communion with all creatures, affirming an eschatological hope that encompasses the entirety of creation in a redeemed future.

A unique contribution of this chapter is the development of an "incarnational pneumatology" to explore how the Spirit actively participates in the redemptive process. I present a vision of the Spirit as dynamically engaged in creation's liberation and as guiding all creatures—not just humans—towards an eschatological future of reconciliation and transformation. This pneumatological soteriology uniquely positions animals as co-creaturely partners within the Spirit's work, recognising them not as passive recipients of grace but as intentional, active participants in the divine drama of redemption.

This innovative approach to animal theology diverges from traditional Christocentric frameworks by emphasising the Spirit's role in uniting all flesh in the redemptive process, echoing prophetic visions of a "new creation" where harmony exists among all creatures. The Spirit's ongoing work is understood to extend beyond humanity, elevating animals to a role within the cosmic restoration that highlights their intrinsic worth and theological significance. In doing so, this incarnational pneumatology reframes salvation as a holistic, inclusive vision that affirms animals' active roles in the eschatological promise of unity and peace. This inclusive vision of salvation, shaped by an incarnational pneumatology, naturally leads to a discussion of eschatology, where the Spirit's transformative work reaches its culmination in the renewal of all creation.

Chapter 5 begins by establishing the theological basis for eschatological hope, emphasising the Spirit's role in uniting personal transformation with the restoration of the natural world. Drawing from perspectives that view eschatology as an opportunity for both spiritual and environmental renewal, the chapter connects the eschatological gift of the Spirit to the transformation of all creation. A central figure in this discussion is Jürgen Moltmann, whose eschatological theology provides a framework for understanding the redemption of animals and the broader ecological community. Moltmann's vision of a "coming Kingdom of God" embraces all flesh, advocating for the resurrection and reconciliation of the natural world. His theology is explicitly animal-inclusive. Animals, as part of God's creation, are not only recipients of divine compassion but active participants in the eschatological vision of justice and restoration. By emphasising the continuity between the present and redeemed creation, this chapter challenges traditional views that separate human salvation from creation's destiny, advocating instead for a holistic eschatology that values all life. The chapter further explores the concept of 're-creation' as a restorative act that honours God's glory. Rather than envisioning a wholly new creation, this view promotes an understanding of salvation as a healing and transformation of the existing world, where animals and other creatures are integral to the eschatological promise. I integrate these eschatological insights into a vision for animal theology, proposing that the eschatological promise of redemption includes the whole of creation. Through the Spirit's redemptive presence, animals are affirmed as co-participants in the divine plan, embodying a pneumatological dignity that resists anthropocentric limitations and envisions a future where all creatures find fulfilment in God's reconciled kingdom.

This reconceptualisation of eschatology as an inclusive theology of hope lays the groundwork for Chapter 6 on Animal Glossolalia, which develops a pneumatological framework that positions animals as Spirit-baptised creatures. Building on the eschatological affirmation of animals as co-participants in God's reconciled kingdom, this chapter explores how animals embody a dignity that transcends anthropocentric limitations through their participation in the Spirit's transformative work. This chapter offers a unique theological framework that brings together animal theology and pneumatology, presenting animals as Spirit-baptised beings endowed with pneumatological dignity. The Spirit's redemptive work not only unfolds within human relationships but reaches outward to include non-human creatures—co-participants in a creation groaning for liberation. Through this inclusive vision, animals are positioned as intentional participants in God's salvific economy, bearing witness to the Spirit's transformative presence, which unites all of creation in a shared hope of renewal and liberation.

In this chapter, I develop the concept of animal glossolalia. I propose that the Spirit's gift of glossolalia, traditionally understood within Pentecostal theology as the gift of tongues, extends beyond human boundaries to encompass all of creation. Building on previous discussions of animals as Spirit-baptised participants in divine communion, I envision a theological framework in which animals partake in a cosmic form of praise, lament, and prayer through their instinctual sighs, groans, and vocalisations. By interpreting these expressions as Spirit-filled communication, I suggest that non-human creatures engage in a unique form of glossolalia, symbolising their deep spiritual connection with the Creator. I introduce this approach by revisiting foundational arguments for animal sentience and spirituality, underscoring the theological argument that animals, as sentient beings, participate in the Spirit's indwelling presence. Drawing on Pentecostal and pneumatological insights, I argue that glossolalia, as a Spirit-inspired form of communication, might also cross species boundaries.

Animal glossolalia is enriched by theological reflections on cosmic worship, notably Richard Bauckham's interpretation of Revelation's "four living creatures," which symbolise a broad spectrum of creation's participation in divine praise. These images suggest that animals and humans alike share in a cosmic chorus of worship, transcending human linguistic limitations. By reimagining glossolalia as a universal Spirit-language, the chapter proposes that animals' vocalisations, however incomprehensible to humans, are woven into the Spirit's intercessory work, serving as unspoken prayers of praise, yearning, and lament.

I then consider the story of Balaam's donkey as a potential Old Testament instance of glossolalic speech, where the Spirit enables a non-human creature to communicate with a human. This narrative illustrates a theological precedent for animal voice and expression, opening the possibility that animals can participate in divine communication under the Spirit's influence. Building on this trajectory, I turn to the concept of *angeloglossy*, inspired by John Poirier's exploration of angelic languages, to argue that glossolalia might also include an esoteric language shared across creation, encompassing both human and non-human voices in the Spirit's cosmic song.

The idea of an animal glossolalia invites readers to reconsider the boundaries of divine communication, urging a more inclusive understanding of glossolalia as a gift accessible to all of creation. By framing animals as Spirit-baptised beings who contribute their unique voices to the worship of the Creator, this chapter employs pneumatological imagination, prompting a recognition of animals as co-participants in creation, contributing their own glossolalic expressions within the broader, Spirit-filled symphony of cosmic worship.

Approach and Method

This study takes shape as a work of literature-based theology, seeking to construct an animal pneumatology within a Trinitarian horizon. In doing so, it not only contributes to the growing field of animal theology but also presses into the often-neglected territory of pneumatological scholarship.

Rather than focusing on a single conversation partner, I employ a thematic approach through the lens of systematic and constructive theology.[2] By engaging various responses to animals within ecumenical and ecological communities, it selectively incorporates perspectives that contribute to a pneumatological understanding of creation. Pentecostal theology provides a consistent thread, strategically positioned to offer a distinctive pneumatological response. This thread serves as a launchpad for broader discussions within Pentecostal and animal theology circles. My conviction is that Pentecostal theology, often overlooked in these discussions, has the potential to raise novel points of dialogue concerning animal ethics, not only within Pentecostalism but ecumenically. I propose that a Pentecostal animal theology holds significant promise in sustaining, creating, healing, and protecting both human and non-human life on Earth.[3] While this endeavour may seem paradoxical to some, it is a deeply

[2] David Clough adopts a thematic approach through the lens of systematic and constructive theology in his work *On Animals*, moving beyond engagement with a single conversation partner to address a broader spectrum of theological and ethical concerns surrounding human-animal relationships. David L. Clough, *On Animals, Systematic Theology* (London: T&T Clark, 2014), xv. Swoboda also adopts a similar method, see A. J. Swoboda, *Tongues and Trees—Towards a Pentecostal Ecological Theology* (Dorset: Deo Publishing, 2013), 9.

[3] A.J. Swoboda supports this perspective in his development of a Pentecostal eco-theology, emphasising the role of the Spirit in sustaining and renewing all of creation, including the non-human world. A J Swoboda, *Tongues and Trees – Towards a Pentecostal Ecological Theology* (Dorset: Deo Publishing, 2013), 9.

Wolfgang Vondey suggests that Pentecostalism offers a distinctive lens through which the Spirit's transformative presence in creation can be understood. Wolfgang Vondey, *Beyond Pentecostalism: The Crisis of Global Christianity and the Renewal of the Theological Agenda* (Grand Rapids, MI: Eerdmans., 2010), 16–46. The imagination, fuelled by Pentecostal narratives of the Spirit's work, enables believers to envision a more holistic approach to healing—a vision that does not separate spiritual redemption from the physical restoration of the earth. Wolfgang Vondey, "Pentecostal Theology," *Journal of Pentecostal Theology* 28 (2019): 44, https://doi.org/10.1163/17455251-02801004. He thus advocates for a Pentecostal engagement with ecological renewal that encompasses environmental stewardship as part of its redemptive mission. This, for Vondey, represents a reclaiming of Pentecostal praxis, where healing is extended beyond individuals to include the restoration of creation itself, thereby integrating theological reflection with practical action aimed at ecological and social justice. The current challenge for charismatic scholars is to expand the understanding of Pentecost by incorporating its imaginative dimensions into both contemporary writing and spiritual testimony. By bridging the divide between current ecological crises and potential futures, and by engaging with narratives that express these visions, Pentecostals are cultivating a more dynamic and actionable pneumatological eco-imagination. The aim of this approach being to address the environmental crisis by integrating the Spirit's transformative work with a forward-thinking ecological ethic.

theologically grounded task that can reshape both Pentecostal thought and wider Christian engagement with the non-human world.

In addition to a thematic, systematic theology approach, I draw upon Amos Yong's model of "pneumatological imagination" to shape my pneumatological consideration of animals.[4] This model serves as an instrument to inform my reading, interpretation, and theological reflection on animals from a pneumato-centric perspective. Shane Clifton summarises Yong's "pneumatological imagination" as a framework that "envisages theology that is biblically grounded, but that reads the Scriptures through a hermeneutical grid informed by the experience of the Spirit in Luke-Acts. In that narrative, Pentecost is not a one-off event, but the template for an open and creative discernment of the Spirit in imaginative interpretations of the Bible and the community; interpretations which look to transform the world."[5] This imaginative approach allows for an expansive reading of the Spirit's work in creation, enabling a fresh theological consideration of animals as participants in the Spirit's activity.

Yong's framework does not remain confined to ecclesial pneumatology but extends into what he calls "foundational pneumatology," which connects the Spirit's presence and activity in the Church with the broader workings of the Spirit in the world.[6] This expanded view provides a robust theological foundation for exploring the role of the Spirit in relation to non-human creatures. By engaging this framework, I aim to highlight how the Spirit's life-giving and transformative presence can be understood as encompassing the entire creation, including animals, thereby grounding a Pentecostal animal theology within a pneumatological imagination.

[4] The concept of *Pneumatological Imagination* is employed as a methodological approach in contemporary Pentecostal scholarship. This approach emphasises the role of the Holy Spirit in fostering creative, imaginative engagement with theological issues, particularly in relation to creation and embodied experiences. See James K. A. Smith, *Thinking in Tongues: Pentecostal Contributions to Christian Philosophy* (Grand Rapids, MI: Eerdmans, 2010), 30.; Nimi Wariboko, *The Pentecostal Principle: Ethical Methodology in New Spirit* (Grand Rapids, MI: Eerdmans., 2012), 38.

[5] Shane Clifton, "Editorial: Identity and the Shape of Pentecostal Theology." *Australasian Pentecostal Studies* 19 (June 15, 2017): 14. See also Yong, *The Spirit Poured Out on All Flesh*, 10; and Amos Yong, *Spirit-Word-Community: Theological Hermeneutics in Trinitarian Perspective* (Eugene, OR: Wipf & Stock Publishers, 2002), 133.

[6] Yong, *Spirit-Word-Community*, 135.

In addition to the primary methodological approaches I employ throughout this book, I adopt an interdisciplinary approach that synthesises findings from scientific inquiry and theological analysis. By engaging with empirical research on animal cognition, sentience, and behaviour, I aim to establish a solid foundation for understanding animals as complex beings capable of emotional and relational depth. This scientific grounding serves as a credible platform for examining themes such as emotional resonance, social behaviours, and intricate modes of communication in animals.

Positioning this Work

The primary field to which my research contributes is animal theology, with a focus on developing a pneumatological perspective on non-human creatures within a Trinitarian framework. This book provides a distinctive contribution to animal theology by systematically incorporating pneumatological insights, an approach that remains largely unexplored. A key innovation of this work is its examination of Spirit-baptism and glossolalia as phenomena that may extend beyond the human realm, thereby challenging traditional anthropocentric theological assumptions.

By expanding pneumatology to encompass non-human creatures, my research raises significant ethical and theological questions about the treatment of animals and their role within God's redemptive plan. This contribution enriches the field of animal theology by proposing a framework that recognises animals as participants in the Spirit's life-giving and transformative work. It moves beyond existing theological frameworks, which often marginalise or overlook animals, and instead offers a robust theological foundation that considers animals as spiritual and ethical subjects with inherent worth.

Additionally, this book advances Pentecostal theological scholarship by extending pneumatology's scope beyond human experiences to include the wider creation. By investigating the Spirit's work in relation to non-human creatures, this research opens new theological pathways for Pentecostal engagement with both ecotheology and animal theology. I suggest that Pentecostal theology, which traditionally centres on human experiences such as Spirit-baptism, can offer a rich framework for understanding the Spirit's work in sustaining, healing, and renewing all of creation, including animals. In doing so, I broaden the scope of Pentecostal thought, positioning it as a valuable resource for ecumenical and interfaith dialogue on creation care.

Moreover, this book contributes to ecotheology by proposing a pneumatological foundation for enhanced ecological responsibility. Recognising the Spirit's sustaining presence in all life, this project provides theological grounding for reimagining Christian ecological ethics, underscoring the interdependence of human and non-human life in the shared mission of creation care. This perspective deepens contemporary discussions on environmental stewardship, encouraging a view of animal protection and environmental responsibility as integral to Christian spiritual and ethical commitments.

In this way, pneumatology becomes a bridge between theology and the pressing concerns of our time, such as animal welfare and the ecological crisis. By engaging in dialogue with environmental ethics, animal studies, and related disciplines, this work opens space for new collaborations. It suggests that the Spirit's activity in creation can inspire fresh ethical imagination and practical responsibility, inviting a more integrated and compassionate approach to life on this fragile earth.

Scope and Boundaries

The theological nature of this book means that it is not grounded in the animal or behavioural sciences—such as zoology, ecology, animal behaviour theories, or natural history in the broader sense—although it may intersect with these disciplines at certain points. The research also does not engage with the practical aspects of animal care, such as conservation biology or wildlife management. While these fields are important and related to the broader context of animal studies, they have been explored elsewhere. In contrast, far less has been written about non-human animals from the perspective of pneumatology within systematic, dogmatic, or constructive theology, which this book aims to address.

Theological reflection, however, does not occur in a vacuum. This work is not intended as a biblical or exegetical study. While a growing body of work in biblical studies has engaged with the theme of animals, this project does not focus on exegesis. Instead, I seek to integrate theological insights from pneumatology into the ongoing conversation within constructive and doctrinal theology. Similarly, while this book is not framed as practical theology, it does have practical implications, particularly concerning how Christians might engage with animal ethics and creation care, which open avenues for further exploration.

While this book provides an in-depth examination of pneumatology in relation to non-human animals, my aim is not to construct an exhaustive systematic theology. Rather, I seek to advance theological reflection in new directions and open a fresh space within systematic theology for considering animals in light of pneumatology. By contributing a systematic approach to this underexplored area, my hope is to inspire further scholarship and deepen the ongoing conversation in animal theology and beyond.

How we Speak of Creatures

Before engaging in a detailed discussion of animal theology, it is important to establish the key terms that will be used throughout this work. In the context of creation theology, words like "animals" and "creatures" are often used interchangeably. However, many scholars prefer the term "non-human animals" to emphasise that humans are biologically animals as well. This linguistic choice aims to bridge the conceptual gap that often separates humans from other animals.[7]

For the sake of clarity and conciseness, I will predominantly use the term "animal" to refer specifically to non-human animals, while "human" will be used to denote human beings. For the sake of variation in language, the term "creature(s)" will also be used. This is not intended as a diminutive term but rather as an alternative to avoid repetition while still maintaining the focus on non-human animals. Importantly, the use of the pronoun "it" will be deliberately avoided when referring to non-human creatures, as this risks depersonalising and objectifying them—something I actively resist. Instead, the focus will be on presenting animals as participants in the Spirit's life-giving work, not as mere objects or stage props within creation.

[7] This definition draws from Clair Linzey's understanding of "animal" and "sentience," where she outlines the importance of recognising non-human animals as sentient beings. I have adapted these definitions to more clearly delineate the focus of this work. See Clair Linzey, *Developing Animal Theology: An Engagement with Leonardo Boff* (London: Routledge, 2022), 1–2.

For the purposes of this book, the term 'animal(s)' refers to non-human animals where sentience is reasonably assumed—such as mammals, reptiles, amphibians, birds, and fish. Sentience, here, refers to the capacity to experience sensations of pleasure and pain, as well as emotional states such as fear, anxiety, and stress. While insects are classified as animals, they are not included in this definition due to the limited yet emerging research regarding their capacity for sentience. The concept of sentience in this work is based on well-established scientific findings from peer-reviewed studies and academic publications. A more in-depth exploration of sentience will be addressed in the next chapter.

Engaging the Conversation

Locating Animals in Pneumatological Reflection

Animals have been a relatively limited aspect of Christian theology and have come into more focused consideration in the twenty-first century. While the Spirit's work in animals has been largely dismissed due to neo-Platonic views of the world,[8] Methodist evangelist and theologian, John Wesley explored this notion in a sermon he preached in 1782.[9] In his sermon, he suggests that God's redemption would be extended to all life, including animals. He elaborates how the eschatological being of every animal would exceed their present ontological state, including their level of intelligence, which would be increased to the human level "as a recompense of what they once suffered."[10]

In 1994, Andrew Linzey established a new frontier in theological discourse by coining the term "animal theology" in his ground-breaking

[8] In the Christian tradition, the ordering of human beings above other creatures has often been substantiated on the understanding that humans are uniquely created the *imago Dei* (Gen 1:26–27). From at least Augustine onwards (in the Western world, particularly), the location of human beings at the top of such hierarchy has been tied to the understanding that the divine likeness is found chiefly in human rationality; creatures without such rationality were thus relegated to the category of 'dumb beasts' (or creatures without sentience). Ian Charles Jones, "Dominion and Communion: Patristic Theology and the Ethics of Humanity's Relationship with Animal Creation." (Ph.D., Fordham University, 2016), 15.

[9] John Wesley, "Sermon 60: The General Deliverance." In *The Works of John Wesley*, 3rd ed., 5: 241–52 (Grand Rapids, MI: Baker Book House, 1991).

[10] Wesley, "Sermon 60," 251.

book of the same name. In this pioneering work, Linzey challenged traditional perspectives by advocating for a theological re-evaluation of the treatment and moral status of animals.[11] Over the subsequent decades, he has authored and edited more than 100 publications focused on the theological implications of human-animal relationships, a contribution unparalleled by any other theologian in the field.[12] Linzey's extensive body of work has cemented his status as a foundational figure in animal theology, inspiring a growing field of scholarship dedicated to understanding animals within a theological framework.

While, surprisingly, Linzey does not himself provide a firm definition for animal theology,[13] his daughter, scholar Clair Linzey, offers

[11] Andrew Linzey, *Animal Theology* (Urbana: University of Illinois Press, 1995).

[12] For an extensive bibliography of relevant works by Linzey see: Oxford Centre for Animal Ethics. "Andrew Linzey." Accessed April 17, 2021. https://www.oxfordanimalethics.com/who-we-are/director/.

[13] See Andrew Linzey, *Animal Gospel* (Louisville: Westminster John Knox Press, 1998), 7–8.

Linzey does offer the following statements which shape a broader, conceptual definition of animal theology:

a. The triune God loves and delights in all creation and creaturely differences.
b. The incarnation signifies that God took on flesh, not just human flesh but creaturely flesh. Flesh is also the basis for sentience.
c. Christ's sacrifice and resurrection on the cross symbolises the liberation of every suffering creature from bondage. Suffering is associated with sentience, as both humans and animals have the ability to feel and respond to pain and suffering.
d. In Christ, humans are commissioned to exercise self-sacrifice. Therefore, humans as a servant-species should serve creation and care for its creatures just as Christ served the weak and vulnerable.
e. The Spirit is the source of life and animates all living creatures. The Spirit leads us towards the peaceable kingdom.

Further to these statements, Linzey makes distinctions between ecotheology and animal theology. It's important to acknowledge these distinctions, as ecotheology encompasses concerns for the entirety of creation, where animal theology focuses specifically on the non-human creature. Linzey delineates these distinctions by offering three specific points. First, Animal Theologians perceive vegetarianism as the initial stride towards forging a world devoid of violence, presenting a contrasting vision of humanity's role within creation. Second, Linzey emphasises the issue of suffering, particularly in the context of hunting, a practice that some ecotheologians do not oppose. Lastly, he examines human intervention in animal species management, where conservationists may prioritise the preservation of endangered species by sacrificing or exterminating others. on the conflict between eco-theology and animal theology see, Andrew Linzey, *Creatures of the Same God: Explorations in Animal Theology* (New York: Lantern Books, 2009), 29–44.

a succinct rendering of the term, explaining, "Animal Theology is an attempt to view the Christian tradition through an animal-friendly lens, while retaining a critical approach to the tradition with regards to its concern for animals."[14]

Linzey's major emphasis could be described as Christological, though he does reference the Spirit; on the subject of the Spirit, Linzey's main arguments can be summarised as containing three key themes:

a. God cannot be indifferent to enspirited creatures who are animated by the life-breath of the Holy Spirit. If non-human animals are "enlivened" by the Spirit, then the sacrificial work of Christ is not restricted to human creatures.[15]
b. The "Spirit-filled" sons and daughters of God are agents of the liberating Spirit to the rest of creation and should model themselves as Christ, "the suffering yet glorified servant."[16]
c. The Spirit opens new possibilities for each generation to live in a violence-free world.[17]

While Linzey brings animals into the fold of Christ's redemptive work through the Spirit, his pneumatology appears subordinated to soteriology. If pneumatology were treated as a distinct or equally weighted locus alongside Christology and soteriology, it could add significant theological depth to Linzey's contributions, particularly regarding the Spirit's role in the broader creation, including non-human creatures.

Linzey clearly emphasises that the role of humans as the servant species as central to "the divine work of redemption to which humans are called by the power of the Spirit." No moral action is possible without the Spirit's enabling presence. It is God who awakens moral insight, instils hope, and, through the Spirit, enables the pursuit of moral action—albeit in a limited way—moving creation toward the realisation of God's kingdom. The references to the Spirit in this context are predominantly human-centered, which may be seen as limiting the broader theological consideration of the Spirit's role in relation to non-human animals. This human-focused lens could overlook the potential for a more expansive pneumatological understanding that includes the ways in which the Spirit might interact with and sustain non-human creatures within the divine order.

[14] Linzey, *Developing Animal Theology*, 3.

[15] Linzey, *Animal Theology*, 25.

[16] Linzey, *Animal Theology*, 72.

[17] Linzey, *Animal Theology*, 89.

Extending the work of Andrew Linzey, David Clough's first volume of the two-volume *On Animals* formally constructs a systematic theology devoted to affirming non-human creatures as sentient, autonomous beings that are worthy of theological attention.[18] Clough examines animals in relation to the doctrines of creation, redemption, and reconciliation to develop an extensive theological framework that highlights "the ways in which humans and other animals stand in the same place before God."[19] Like Linzey, Clough emphasises a Christocentric responsibility to actively contribute to the flourishing of non-human creatures as an act of Christ-imitating service to the vulnerable. However, a limitation in Clough's approach is his lack of engagement with the doctrine of the Holy Spirit, which could further enrich and extend his systematic theology of animals. Building on Clough's Christocentric approach, my research aims to extend the conversation by incorporating pneumatology—highlighting the essential role of the Holy Spirit in the flourishing and sustenance of all creatures. By integrating pneumatology into animal theology, I seek to develop a more comprehensive framework that acknowledges the Spirit's active role in the lives of non-human animals.

The most expansive pneumatological consideration of animals (within the field of animal theology) to date can be attributed to Ryan Patrick McLaughlin. McLaughlin's primary focus is not on constructing a pneumatology, yet his engagement with Moltmann naturally integrates pneumatology from an eschatological perspective. McLaughlin aims to develop an animal ethic that aligns with Moltmann's theological vision by bringing animal theology into dialogue with Moltmann's broader framework. Although McLaughlin's study does not centre on pneumatology, he explores a range of topics, including pneumatology, for their potential contributions to a "maximally inclusive eschaton" for non-human

[18] See David L Clough, *On Animals, Systematic Theology*, Vol. 1 (London: T&T Clark, 2014) and David L Clough, *On Animals, Theological Ethics*. Vol. 2 (London: T&T Clark, 2019).

[19] Clough, *On Animals, Theological Ethics*, 44.

creatures. In[20] this context, McLaughlin synthesises Moltmann's pneumatology into three key points to propose relevant insights for animal theology:

a. The breath (*ruach*) of Elohim is the principle of life present in all living creatures through the Spirit.[21]
b. As the Spirit's manifestation is present and active in the sighs and groans of the cosmos, the Spirit is also preserving the cosmos as *creatio continua*.[22]
c. While the Spirit preserves life, the Spirit also sustains life as *creatio nova*, leading the cosmos into its transfiguration and eschatological future.[23]

McLaughlin's engagement with Moltmann's cosmic theology unsurprisingly leads to conclusions rooted in eschatology. First, he draws on Moltmann's panentheistic vision, asserting that all creatures are included in the "all flesh" destined for resurrection.[24] Second, McLaughlin highlights Moltmann's concept of a cosmic community, one that necessitates "laws to safeguard the integrity of its members."[25] From these insights, McLaughlin argues that if animals are truly partners with humans as members of "all flesh," they are recognised before God as beings of intrinsic worth. This recognition implies that animals are entitled to the same rights as all members of the cosmic community, sharing in the blessings of cosmic transfiguration.[26]

McLaughlin's interpretation of pneumatology is particularly significant, as it extends the Spirit's work beyond humanity to include all life. This eschatological framework offers a valuable resource for animal

[20] Ryan Patrick McLaughlin, "Anticipating a Maximally Inclusive Eschaton: Jürgen Moltmann's Potential Contribution to Animal Theology." *Journal of Animal Ethics* 4, no. 1 (2014): 19. McLaughlin explores Moltmann's doctrines of the Trinity, Pneumatology and Eschatology and others.

[21] McLaughlin, "Anticipating a Maximally Inclusive Eschaton," 25.

[22] McLaughlin, 25.

[23] McLaughlin, 25.

[24] McLaughlin, 26.

[25] McLaughlin, 30.

[26] McLaughlin, 30.

theology, suggesting new possibilities for understanding the Spirit's redemptive activity in a more inclusive and creation-wide scope. His work points towards a pneumatology that aligns well with Pentecostal perspectives, suggesting further avenues for exploring the Spirit's eschatological role in the redemption of non-human creatures.[27] Yet, despite McLaughlin's contributions—alongside those of other animal theologians—there remains a noticeable gap. Extensive and sustained engagement with pneumatology as it relates to animal theology is still largely underdeveloped.

Broader Contemporary Approaches

In *Breath of Life*, Australian theologian Denis Edwards develops a comprehensive theology on the Creator Spirit that affirms the value and significance of non-human animals within the broader scope of creation. While his theology extends to all of creation, it provides substantial anchor points regarding the intrinsic worth of animals, which will be referenced throughout the body of this research. Drawing upon rich imagery of the Spirit as breath, wind, oil, and fire, Edwards evokes a sense of God's presence that transcends human experience. He asserts that the Creator Spirit is in intimate communion with every living creature, each one "loved into being" and drawn towards a future with God.[28] Like Irenaeus before him, Edwards contends that non-human creatures have always been part of this divine communion through the Spirit, who continually breathes life into creation and sustains all things in existence.[29] The Spirit's presence is not confined to humanity but flows freely through the natural world, fostering a web of life characterised by mutuality and interdependence.

Building on this pneumatological tradition, Mark Wallace offers a broadened understanding of the Spirit's embodiment within the Christian faith. He argues that Christianity embodies a "double incarnation," in which God becomes flesh not only in human beings but also in non-human entities. Wallace writes, "just as God became human in

[27] As Frank Macchia explains, "Eschatology can only involve all of creation in its implicit yearning for the kingdom to come if the Spirit is seen as involved in the reach of life toward renewal in all things." Frank D Macchia, *Baptised in the Spirit: A Global Pentecostal Theology* (Grand Rapids, MI: Zondervan, 2006), 41.

[28] Denis Edwards, *Breath of Life—A Theology of Creator Spirit* (Maryknoll, NY: Orbis Books, 2004), 114.

[29] Edwards, 114.

Jesus, thereby signalling that human beings are the enfleshment of God's presence, so also becoming avian in the Spirit, God signals that other-than-human beings are also the realisation of God's presence."[30] Wallace's insight expands the traditional view of incarnation by including non-human creatures in the divine embodiment, specifically the avian form symbolised by the dove in biblical creation narratives. This expansion suggests that non-human beings also carry the presence of God, participating in the same divine reality as human beings.[31]

Wallace furthers this notion by suggesting that Christian animism—an ancient belief in the spiritual presence in all creatures—naturally aligns with a Christian pneumatological framework.[32] This view does not stand in tension with Christianity but instead reveals its intrinsic spiritual affinity with the natural world—where the Spirit of God breathes through all forms of life. Wallace's interpretation calls for a recognition of animals and the broader natural world as sacred participants in the unfolding of God's Spirit, urging a theological shift that moves beyond anthropocentrism.

Lutheran theologian, Wolfhart Pannenberg offers a generous pneumatological view that affirms non-human creatures are integral to God's good creation. In his systematic theology, Pannenberg makes key reference to the Spirit's animating presence within non-human creatures. While his consideration may have humans primarily in view, Pannenberg upholds that the "life, movement and activity" of non-human creatures can be attributed to the Spirit of God.[33] For Pannenberg, the Spirit is the integral creative presence of God within creation, as the Spirit animates

[30] Mark I. Wallace, *When God Was a Bird: Christianity, Animism, and the Re-Enchantment of the World* (New York: Fordham University Press, 2019), 85.

[31] Mark Wallace's perspective on the Spirit highlights a neglected model in Western theology, which views the Spirit as a life force that is closely tied to nature. Although in theory, the Spirit has always been described as both the Spirit of God and the Spirit of creation, with the former embodying the power of reciprocity within the Trinity and the inner force of human redemption, and the latter as the breath of God that animates and sustains the universe, in practice the emphasis has almost exclusively been on the Spirit of God. As a result, the cosmic role of the Spirit as the life-giving force within all of creation, both human and non-human, has been consistently overlooked. Mark I. Wallace, *Fragments of the Spirit: Nature, Violence and the Renewal of Creation* (New York: Continuum, 1996), 136.

[32] Wallace, *When God Was a Bird*, 85.

[33] Wolfhart Pannenberg, *Systematic Theology Vol 2*. Translated by Geoffrey W. Bromiley (London: T&T Clark, 2004), 76.

life and unites all creatures in fellowship with their Creator.[34] The beauty of Pannenberg's pneumatology is that he does not divorce the Spirit from the creaturely realm; rather, his account of the 'continuous creation'attests to the giving and sustaining presence of the Spirit in all of created life.[35]

Like Moltmann, Pannenberg, in his robust theology, welcomes the integration of pneumatology and eschatology, where the Spirit's action in the present provides hope for the eschatological fulfilment of all creation. He explains that while an eschatological consummation remains unfulfilled, life in the here and now is a proleptic expression that provides all creatures with hope for the future new creation.[36]

Pannenberg's proleptic eschatology provides a platform for Macchia's work in human glossolalia, where glossolalia is expressed in anticipation of our eschatological orientation, as "a foretaste of eschatological transcendence."[37] The Spirit's prompting of creation's sighs and groans for fulfilment as described in Romans 6:26 illustrates this prolepsis, whereby glossolalia is an expression in the here and now. I will now contextualise this notion to non-human creatures. As glossolalia points to eschatological fulfilment and if animals are, speculatively, glossolalic participants, a shadow of their fuller eschatological orientation manifests in the present through God's people (through human stewardship), and the peaceable Kingdom depicted in Isaiah 11:6 portrays the future relationship between animals, humans, and God in the new creation.

Therefore, if eschatology is the context for Spirit-baptism, which prepares all of creation for a shared eschaton, then the implications are a shared experience of creaturely unity—reflective of the "all flesh" of the Spirit's baptising work. I will explore this notion further as I turn to a brief survey of Pentecostal scholarship.

[34] Pannenberg, *Systematic Theology Vol 2*, 32.

[35] Pannenberg, *Systematic Theology Vol 2*, 76. Pannenberg's account of continuous creation refers to the 'preservation' of creation, where creation is not a one-time historical act but a continual work of God in the Spirit.

[36] Wolfhart Pannenberg, *Systematic Theology Vol 3*. Translated by Geoffrey W. Bromiley (London: T&T Clark, 2004), 553.

[37] Macchia. *Baptised in the Spirit*, 281.

The Pentecostal Perspective

With a characteristic emphasis on pneumatology, Pentecostals are well-positioned to engage pneumatologically with animal theology and broaden this area of research by constructing a more robust pneumatological animal theology.[38] Unfortunately, to date, there are few Pentecostal scholars who have addressed the theological place of animals in relation to the Spirit. As Pentecostal resources on animals are limited, this research will explore Pentecostal considerations of creation care and ecopneumatology. In the process of assessing these two categories, I will emphasise any reference made by Pentecostal scholars to the status of non-human creatures within a pneumatological context.

One Pentecostal theologian who has begun to address this gap is Amos Yong, whose work provides several key points in relation to animals within his broader "pneumatological theology of creation." Yong outlines three significant insights that inform his approach to non-human creatures. First, in *The Spirit of Creation*, he emphasises that while the *ruach Elohim* enlivens non-human animals, these beings are not classified as moral agents.[39] This distinction is important because it highlights the theological boundary that has traditionally separated human beings from other creatures in terms of moral agency. Second, Yong briefly acknowledges the sentience of animals, noting that they possess a capacity for experience and feeling, though this observation is not developed in great detail. Finally, in *The Missio Spiritus*, Yong underscores the interrelatedness between humans and non-human animals through the work of the Spirit, suggesting a deeper, pneumatological connection across creation.[40]

In his Pentecostal reading of Genesis 1, Yong highlights the cooperation between *ruach Elohim* and the natural world in the processes of creation. For example, he notes that while fish and birds are "enspirited creatures" animated by the Spirit's breath of life (Gen 1:30), they are

[38] Denis Edwards acknowledges that while his work on the Creator Spirit is extensive, there is an area of limitation—namely, the lack of engagement with Pentecostal literature on the Spirit. Edwards himself notes that this is an underexplored aspect within his broader theological reflections on pneumatology. See Edwards, *Breath of Life*, 194n.1.

[39] Amos Yong, *The Spirit of Creation: Modern Science and Divine Action in the Pentecostal- Charismatic Imagination* (Grand Rapids, MI: Eerdmans, 2011), 160 n 67.

[40] See Amos Yong, "The Missio Spiritus: Towards a Pneumatological Missiology of Creation." In *Creation Care in Christian Mission*, ed. Kapya Koama (Vol. 29, 2015), 126. See also Yong, *The Spirit of Creation* in its entirety.

not ontologically considered moral agents.[41] This distinction is significant within Yong's pneumatological framework, which opens up theological space for considering animals as part of the Spirit's life-giving work, yet without moral accountability in the human sense. However, despite Yong's framing of a "pneumatological theology of creation," his engagement with animals remains relatively brief and leaves room for further exploration.

While Yong offers a foundation for considering the Spirit's role in the life of non-human creatures, his work highlights the need for more sustained engagement within Pentecostal theology. As the following literature review will demonstrate, Pentecostal scholarship remains underdeveloped in terms of fully integrating non-human animals into a pneumatological framework. This gap provides an opportunity for further theological reflection on the Spirit's work in sustaining and renewing all life, human and non-human alike.

In the same work, Yong considers how the complexity of human beings involves the ability to share experiences, demonstrate higher levels of individuality, spontaneity, and self-determination—characteristics that are dissimilar to rocks yet shared by "other sentient beings (like animals)."[42] While he acknowledges animal sentience here, he does not further develop this notion, much like his earlier treatment of *ruach Elohim* for "lower-level" entities.[43] However, when reflecting eschatologically in another publication, Yong invokes Isaiah's pneumatological vision of the peaceable kingdom, where "the Spirit is poured out upon all flesh, including the wolf and the lamb, the leopard and the kid."[44] This eschatological framework hints at a future where the Spirit's work reconciles all of creation, human and non-human alike.

Moreover, in *The Missio Spiritus*, Yong reiterates the theme of interdependence, emphasising that "humans are pneumatologically interrelated not only with one another but also with non-human animals since all of

[41] Yong, *The Spirit of Creation*, 160 n 67.

[42] Yong, *The Spirit of Creation*, 200.

[43] Amos Yong, *Renewing Christian Theology: Systematics for a Global Christianity* (Waco, TX: Baylor University Press, 2014), 283.

[44] Amos Yong, *The Spirit Poured Out on All Flesh: Pentecostalism and the Possibility of Global Theology* (Grand Rapids, MI: Baker Academic, 2005), 300.

life throbs with and through the breath given by the *ruach* of God."[45] He suggests that the *ruach Elohim* creates, sustains, and thus links all life together, while maintaining levels of subordination dictated by varying degrees of sentience.

Yong's pneumatological framework resonates with McLaughlin's theological structure, which builds on Moltmann's view that all life exists within a pneumatological context. While Yong's exploration of the subject is valuable, it remains fragmented and does not cohesively address the subject of animals in one comprehensive body of work.[46] Nevertheless, his contributions offer a pneumatological foundation for further exploring the ontology of non-human creatures and the reconciliation between human and animal life within a Pentecostal framework. Yong's scholarship on the subject of animals, while helpful, does not expand further than what is summarised here, nor does it appear cohesively in one body of work.[47] Yet, while Yong's contribution is limited, it does provide

[45] Yong, "The Missio Spiritus," 127.

[46] In other areas of Amos Yong's scholarship, animals have been discussed, though not always through the lens of pneumatology. For instance, Yong engages with theological perspectives on animals in relation to sin, death, and redemption, particularly concerning how sin, through Adam, impacts not only humans but also animals. This discussion underscores the broader theological narrative of human fallenness and the suffering experienced by non-human creatures. While these reflections are crucial for understanding animals' role in eschatological redemption, they do not necessarily delve deeply into pneumatological dimensions, focusing instead on broader spiritual frameworks. See Yong's section on Creation and the Fall in Yong, *Renewing Christian Theology*, 259–291. For Davis, Yong advances the way forward by advocating for an alternative reading through the lens of the Spirit's soteriological work in creation. She notes that Yong acknowledges that any renewed doctrine of creation must grapple with coherent explanations of suffering, death, and sin. Addressing these hermeneutical issues is essential for overcoming barriers posed by literalism and fostering a more constructive engagement with both evolutionary science and ecological concerns. Anita Davis, "Pentecostal Approaches to Ecotheology: Reviewing the Literature," *Australasian Pentecostal Studies* 22, no. 1 (June 8, 2021): 16. While I critique Amos Yong's limited engagement with animals in his work, I acknowledge his pioneering contributions to the broader ecotheological discourse through a Pentecostal lens. See Amos Yong, Eugene Baron and Amos Yong, eds., *Pentecostal Missiology and Environmental Degradation* (Carlisle, UK: Langham Global, Forthcoming); Amos Yong, ed., *The Spirit Renews the Face of the Earth: Pentecostal Forays in Science and Theology of Creation* (Oregon: Pickwick Publications, 2009).

[47] In a forthcoming publication on eco-translation, Amos Yong acknowledges that his earlier theological work has been primarily anthropocentric in focus. This recognition represents a shift in his scholarship as he begins to engage more directly with non-human creation. While his prior writings haven't addressed typical aspects of animal theology,

pneumatological scope to explore an ontology of non-human creatures which works to reconcile human life to animal life within a Pentecostal framework.

In Pentecostal eco-theology, the connection between Spirit baptism and ecological mission is becoming increasingly significant. Scholars like A. J. Swoboda and Amos Yong have explored how pneumatology can provide fresh insights into the broader creation. Swoboda introduces the concept of *Eco-Glossolalia*, a metaphor for a Spirit-baptised creation, envisioning a world where all creatures, human and non-human, are united through the Spirit's empowering presence.[48] His interpretation of Pentecostal theology embeds ecological concerns into a Spirit-filled worldview, viewing creation and the church as holistic, interconnected bodies.[49] As with the work of many ecotheologians, the limitation of Swoboda's Pentecostal ecotheology lies in its generalisation of creation, which broadly encompasses the natural world and its creatures without specifically and comprehensively addressing non-human animals as sentient beings. This approach overlooks a distinct ontological consideration of non-human creatures, one that would integrate human and animal life through the

Yong himself now reflects on the limitations of his earlier framework, signalling a movement towards a more inclusive pneumatological vision that encompasses both human and non-human creatures. Amos Yong, "Translation as Social and Ecological Justice (A Review Essay): Notes toward a Theology of Tongues and Interpretations in the Age of the Anthropocene," *Pneuma*, Forthcoming (2025).

[48] A J. Swoboda, "Eco-Glossolalia: Emerging Twenty-First Century Pentecostal and Charismatic Ecotheology." *Rural Theology* 9, no. 2 (2011): 103; Swoboda's metaphor is constructed on the communal understanding of Pentecost (as described in Acts 2:17), where the Spirit baptises "all flesh." Swoboda clarifies that his thesis of a Spirit-baptised creation is developed from the creational narratives of Genesis 1–3, a portrait "of God's Spirit giving life and taking life in every element of the created realm." The tension, as Swoboda notes, is that the fulness of Spirit Baptism is reserved to the Christian (human) experience and not extended to animals. A J Swoboda, *Tongues and Trees: Towards a Pentecostal Ecological Theology* (Dorset: Deo Publishing, 2013), 240–241.

[49] Swoboda, *Tongues and Trees*, 241. While Swoboda explains that Spirit baptism is unique to Christian experience, he does not entirely disqualify a shared experience by non-human agents, which he explains, cannot yet be "discerned." It is within this area of grey where I endeavour to locate my research. Swoboda. *Tongues and Trees*, 240. As referenced earlier, Amos Yong expands on this by referencing the Prophet Isaiah's vision of the Spirit poured out on all flesh, not only human but non-human creatures as well. Yong's approach underscores the idea that the reconciliation and healing of creation are central to the Spirit's work, including animals like the wolf and the lamb in the eschatological promise of peace and renewal. Yong, *The Spirit Poured Out on All Flesh*, 300.

Spirit in a reconciliatory framework. Nevertheless, the strength of Swoboda's research is evident in its notable departure from an anthropocentric perspective. He advocates for a "charismatic theology of creation [that] seeks to heal broken relationships between humans and between human and non-human creation," marking a significant shift towards a more inclusive theological vision.[50]

Frank Macchia also contributes by discussing how creation groans for liberation through the Spirit.However, his emphasis remains on human cultural participation in creation's renewal, leaving the specific role of animals less explored.[51] A pneumatology of animals could bridge this gap, offering a more inclusive vision of how the Spirit's work of transformation applies to all of creation, not just humanity.

Adding caution to this exploration, Veli-Matti Kärkkäinen warns that theological reflection on animals should not stray too far from biblical traditions.[52] He affirms the Spirit's role in animating all creatures, yet his reflections on animals remain limited.[53]

While limited, these considerations of non-human animals by Yong and Kärkkäinen provide a starting point and, perhaps, a caution to explore the spiritual and ontological place of animals through Pentecostal lenses. The references to animal sentience and the interrelatedness of non-human and human animals via *ruach Elohim* are particularly pertinent for the research proposed in this paper, providing Pentecostal scholarship with pathways for further pneumatological exploration.

[50] Swoboda, *Tongues and Trees*, 214.

[51] Frank D Macchia, "The Spirit of Life: Toward a Creation Pneumatology." In *Third Article Theology: A Pneumatological Dogmatics* (Minneapolis, MN: Fortress Press, 2016), 129.

[52] Kärkkäinen, *Creation and Humanity*, 176. Kärkkäinen asserts that while an eschatological and soteriological vision encompassing all creation is crucial, it should not be used as a reason to diminish or overlook humanity's unique role within God's plan for salvation. He emphasises that the theological significance of humanity must be maintained, even within the broader context of creation's redemption.

[53] Veli-Matti Kärkkäinen, *Toward a Pneumatological Theology: Pentecostal and Ecumenical Perspectives on Ecclesiology, Soteriology, and Theology of Mission* (Maryland: University Press of America, 2002), 226.

Conclusion

A pneumatological framework invites us to reimagine the theological place of animals—not as passive elements within creation, but as participants in the Spirit's life. While animal theology has rightly affirmed the value of non-human creatures, the pneumatological dimension of that conversation remains largely unexplored.

By recognising animals as spiritual agents with their own unique relation to the Spirit, this work challenges anthropocentric assumptions and extends the horizons of both pneumatology and animal theology. The implications are far-reaching: to see animals in this way is to discern them not merely as recipients of God's creative breath, but as beings with intrinsic worth who share in the Spirit's sustaining, renewing, and life-giving work.

My hope is to deepen theological reflection on animals while also encouraging a more inclusive and compassionate ethic—one that honours the dignity of our fellow creatures and takes seriously their participation in God's world. The next chapter turns to the ontology of animals and their place within the divine order, exploring how their value and purpose might be understood through a Spirit-centred lens. In this way, the argument advances a pneumatological approach to animal theology that speaks not only to theological discourse but also to Christian ethical practice.

CHAPTER 2

Creation and Covenant

I turn to the pneumatological dimensions of creation, beginning with the breath that animates every living creature. The biblical language of *ruach* and *nephesh* (so often rendered as "breath," "soul," or "life") carries a richness that resists reduction to mere biology.[1] These terms point to the Spirit's presence woven into the fabric of creation, a presence that moves through the animal kingdom as more than simple preservation. What emerges here is a more expansive vision: animals not as passive products of divine breath but as Spirit-imbued creatures, bearing purpose, presence, and participation in God's ongoing creative life.

Animals, as recipients of the Spirit's breath, are envisioned as beings endowed with an ontological capacity that connects them to a relational purpose within creation. This suggests that the Spirit's creative impulse does not simply sustain animal life but rather empowers these creatures to respond to God in ways that are unique to their nature. Such a reading

[1] I intentionally avoid framing this argument around the concept of 'soul' often associated with anthropocentric theological constructs. Instead, I emphasise the Spirit (*ruach*) as the unifying and animating presence that sustains all creaturely life. This pneumatological perspective shifts the focus from the brain or cognitive capacities, often seen as markers of value or distinction, to the Spirit as the source of vitality, communication, and empowerment across the entirety of creation. Such an approach aligns with the broader theological aim of overcoming dualistic paradigms and foregrounds as the unifying thread in all her creatures. For deeper discussions regarding the association of animals and the concept of the soul, see Linzey, *Why Animal Suffering Matters*, 25–29.

D. Rizzo, *Animal Pneumatology*, The Palgrave Macmillan Animal Ethics Series, https://doi.org/10.1007/978-3-032-18723-9_2

challenges anthropocentric perspectives that restrict spiritual capacity to humanity alone, advocating instead for an inclusive framework where all creatures participate in God's covenantal relationship with creation.

As I will explore later in this chapter, emerging research and theological considerations reveal that animals may exhibit forms of spirituality that do not depend on human-like cognitive functions, but rather on affective, relational, and extra-linguistic expressions. The behaviours observed in various animal species, such as the ritualistic practices documented by Jane Goodall and Marc Bekoff, indicate that animals may possess a capacity for awe, mourning, and even proto-religious expressions. These behaviours suggest that animal spirituality may emerge from a shared physiological basis in brain regions associated with emotional and affective experiences. Recognising these dimensions expands our understanding of how animals might experience and express a spiritual life that resonates with the Spirit's creative and sustaining work.

In the pages that follow, I explore how the Spirit's gift of *ruach* and *nephesh* is bestowed upon non-human life, and consider the ontological weight of these terms as they open fresh insight into animal sentience and its capacity for spiritual life.

Yet the creation account does not speak of Spirit alone: it also gestures towards the Word through whom all things came into being. "In the beginning," God's speech summons light from darkness and calls forth the teeming life of earth, sea, and sky. The New Testament identifies this creative Word with Christ himself (John 1:1–5; Col 1:15–18; Heb 1:1–3). In this light, animals come into existence not only by the Spirit's gift of *ruach* and *nephesh*, but also through the *Logos* who is their ground and goal. Just as the Spirit hovers over the primordial waters, the Word orders creation and directs it towards its fulfilment. For non-human creatures, this means their very breath, relationality, and vitality are Spirit-filled yet also Christ-shaped, sustained in the present and destined for renewal in him. Luke's account of the incarnation (Luke 1:35) confirms this inseparability: the Spirit's overshadowing and the Word's enfleshment together mark the beginning of creation's renewal, encompassing animals as well as humans. To read Genesis with both Spirit and Word in view is therefore to see animals not as incidental, but as participants in a creation held together, enlivened, and ultimately healed by God's Triune life.[2]

[2] I am grateful to Frank D. Macchia for encouraging me to bring Christology into closer conversation with pneumatology in relation to the creation account.

In presenting animals as capable of a relational response to the divine, this chapter seeks to enrich our theological vision of creation. It points to a Spirit-filled life that is purposeful, relationally dynamic, and spiritually significant. Drawing on scriptural witness, such as Psalm 148's summons for all creation to praise God, it affirms that the Spirit sustains every creature in a fellowship with the Creator. In doing so, it broadens our sense of spiritual participation beyond the bounds of humanity.

While Clough's systematic theology provides important foundations for this study, this chapter both builds upon and departs from his framework. It does so by placing stronger emphasis on the Spirit (a dimension Clough engages only lightly) and by drawing attention to the idea of the "animal voice." In exploring the Spirit's intercessory role on behalf of non-human creatures, as well as the possibility that animals may express spirituality through non-conceptual forms, I seek to show the Spirit's active work in sustaining and uplifting life within the animal kingdom.

Life Force and Breath

I begin by exploring the theological significance of *ruach* in relation to animals, drawing heavily on the insights of John Levison, whose scholarship offers a nuanced understanding of the Spirit's indwelling presence across all creation.[3] Levison's work, particularly in *Filled with the Spirit*, challenges traditional interpretations that limit the Spirit's activity to human beings or specific moments of empowerment.[4] Instead, he presents a vision of the Spirit as a continuous, life-sustaining force that dwells within all living creatures, human and non-human alike. This view aligns with the Yahwist and Priestly creation accounts, where animals and humans share the same divine breath, suggesting a theological continuity across species. Levison's careful textual analysis—especially his reframing of the Spirit not as an occasional endowment but as an ongoing presence—provides a foundational framework for understanding animals as active participants in God's Spirit-filled creation. By drawing on Levison, I seek to highlight the biblical basis for regarding animals as more than mere living beings; they are recipients of God's Spirit, sharing in the same

[3] Throughout this book, *ruach* and *nephesh* will be used in transliterated form.

[4] Levison, *Filled with the Spirit*, 26.

breath that animates all life. My approach not only advocates for the theological dignity of animals but also reinforces a holistic view of creation as deeply interconnected through the Spirit.

Furthermore, while the concepts of *nephesh* and *ruach* have been explored in relation to animals within pneumatological scholarship, I limit the scope in this chapter to specific theological considerations that underscore the Spirit's activity across creation. By focusing on non-anthropocentric texts that explicitly reference animals, this chapter aims to broaden pneumatology to encompass the full expanse of God's creation, positioning animals as integral to the life-sustaining work of the Spirit. This approach is essential to advancing a theology that honours the value of animals and recognises the Spirit's role in sustaining all creaturely life.

Nephesh and Ruach

The opening of the creation narrative presents the Spirit of God as a dynamic force that infuses life into the world. This breath, or *ruach*, embodies God's creative and covenantal love. "*Nephesh*" and "*ruach*" are elements that transcend the binary distinction between humans and animals in the Old Testament. *Ruach* operates indiscriminately across all living creatures, where the vital essence is not bound by discriminatory boundaries but instead pervades all creaturely life.

Amos Yong's pneumatological reading of creation identifies the theological depth of *ruach* as both breath and spirit, beginning with the priestly account in Genesis, where the *ruachElohim* (often understood as the breath, wind, or even storm of God) moves over the primordial waters (Gen. 1:2). In the second creation account, the Spirit's life-giving breath animates Adam, formed from the dust, illustrating the Spirit's transformative power to bring life to barren places. Here, *nephesh*, often associated with breath, is granted to all living beings (Gen. 1:30), revealing a shared vitality between humans and animals.[5] *Ruach*, commonly translated as

[5] See Jensen, "Discerning the Spirit," 3. Amos Yong, *The Cosmic Breath: Spirit and Nature in the Christianity-Buddhism-Science Trialogue* (Leiden; Boston: Brill, 2012), 60. Yong notes that it is important to approach this interpretation with caution. While it is possible to make connections between the "breath of life" and the divine spirit, it is not necessarily appropriate to impose a specific trinitarian understanding onto this ancient text. For Yong, reading the *ruachElohim* as a reference to the third person of the Trinity might be anachronistic. The Yahwist narrative suggests a unified ontology between humans and animals, as both are animated by the same life force and are described as *nephesh*

"spirit," also conveys breath and wind, suggesting the Spirit's dynamic, uncontainable nature, always reaching outward into the world.[6] Levison adds that the life-giving characteristics seen in Adam's creation (Gen. 2:7) extend to all creatures, implying that animals are not mere objects but beings with a relational connection to God similar to that of humans.[7]

In the Hebrew Bible, *nephesh* is a term often translated as "soul," yet its meaning extends beyond the human-centric notion of a disembodied spirit. It signifies "living being" or "life force" and applies broadly to both humans and animals, blurring the line between them in terms of shared

hayyah—living beings. No explicit ontological distinction is made between human and non-human creatures in their relationship to *adamah* (soil), reflecting a shared earthly origin and destiny. The term *adam* (human) shares an etymological root with *adamah* (soil), underscoring this connection. Mortality, in this framework, is symbolised by a return to the earth, emphasising the cycle of life sustained by *ruach*. The "breath of life" (Gen. 7:22), then, is not an exclusive endowment upon humans but a corporeal breath shared by all animate beings, highlighting that *ruach* serves as a universal life force linking all *nephesh* to God's sustaining power. Hiebert, *The Yahwist's Landscape: Nature and Religion in Early Israel*, 63. Ken Stone also comments on these distinctions. See Stone, *Reading the Hebrew Bible with Animal Studies*, 38.

[6] The term "Holy Spirit" is used only twice in the Old Testament: Ps 51 and Is 63. In the Old Testament, the term 'Spirit' is used in various ways, sometimes to refer to God's activity in the world, other times to express the dynamism of the human creature, and sometimes to imply an interaction between divine spirit and the spirit of God's creatures.

[7] Levison writes, "animals live in God's presence, possess God's spirit, and return to the dust." Levison, *Filled with the Spirit*, 26. Also, see Eccl 3:18–20, "I said in my heart with regard to *the sons of adam* that God is testing them to show that they are but animals. For the fate of the sons of Adam and the fate of the animal is the same fate; As one dies, so dies the other. They all have the same spirit, and the *adam* has no advantage over the animal; for all is vanity. All go to the one place; All are from the dust, and all turned to dust again." The dimensions of the animal creation described is this verse are not exclusive to humans, unlike what is depicted in the creation narratives in Genesis. The Spirit's presence extends to the animal realm, and just like humans, animals are given the Spirit within, which invigorates them with essential and physical vitality.

Levison provides additional insights into the Biblical understanding of the Spirit of God and its connection to all living creatures. He argues that the phrase 'the Spirit of the Lord' is better translated as "a spirit of the Lord," suggesting that the Spirit's presence is not a special, momentary endowment but a lifelong, accessible indwelling. This interpretation reframes being "filled with the Spirit" as a continuous state rather than an isolated or extraordinary event, inviting a re-evaluation of traditional readings of the Spirit's role in biblical narratives. Levison, 33.

life essence.[8] The term *nepes hayya* is employed for both human creatures and non-human creatures, and neither is ascribed a distinct soul or spiritual entity apart from their physical existence.[9] While humans possess unique qualities, they remain fundamentally part of the broader community of living beings created on the sixth day, sharing their origin in dust with non-human creation and their infusion of Spirit with the divine.[10] The usage of *nephesh* signifies embodied life and relational existence rather than an individualised, isolated soul, placing humans and animals within an interconnected creation.[11] The concept of the soul is often viewed through the lens of Stoic philosophy as the rational component of human nature. Rather than engaging with the complexities of that philosophical discourse, my aim is to explore alternative ontological qualities that might be ascribed to animals. Levison draws attention to Philo, who leaves no ambiguity regarding the relationship between inspiration and creation. Philo associates *nephesh* with the purest form of the soul, distinct from the soul entangled with a body susceptible to moral weakness. This illustrates how a first-century thinker could interpret the inspiration in Genesis 2:7 not merely as the act of breathing life but as a profound infusion of divine

[8] Stone, *Reading the Hebrew Bible with Animal Studies*, 37.; Garvey, *God's Good Earth: The Case for An Unfallen Creation.*, 33. The creation narratives employ the term 'שֶׁפֶנ' (*nephesh*) not only in reference to humans but also to animals, binding them together in a shared essence of life. In the Hebrew Scriptures, blood is the sign of that life-force, pulsing through every creature alike. Even in Genesis 9, when permission is given for humans to eat other animals, the blood is withheld, marked as sacred, a reminder that life itself belongs to God. The text's portrayal of humans and land animals being formed on the same day gestures toward their nearness and connection, a vision of kinship woven into the very fabric of creation.

[9] Hiebert, *The Yahwist's Landscape* 63. Hiebert notes that this interpretation has endured for centuries in English translations, as translators often sought to draw distinctions between humans and animals that do not inherently exist in the Hebrew text.

[10] Drawing from an Indigenous perspective, Native American theologian Randy Woodley, interprets the Genesis narrative of Adam naming the animals as an invitation to relationality rather than dominion. Woodley explains that, in an Indigenous context, naming requires a deep knowledge of the one being named, encompassing an understanding of their unique characteristics. Thus, God's instruction for Adam to name the animals is, in Woodley's view, a call for Adam to engage with his "relatives" in creation. This reading aligns the Genesis creation story with Indigenous creation narratives, both of which view humans and the natural world as part of "the same family tree in the heavens and the Earth." Woodley, *Indigenous Theology and the Western Worldview*, 57.

[11] Levison, *Filled with the Spirit*, 148.

essence. Philo's connection of the breath in Genesis 2:7 with the uncreated image or *logos* in Genesis 1:27 enables him to argue that the soul, as a portion of this uncreated breath or image, is itself inherently uncreated. This interpretation sets the divine Spirit within the Torah apart from the Stoic notion of spirit, presenting a uniquely elevated view of divine inspiration. I propose that while humans may face challenges in preserving their *nephesh*—as illustrated in Gen 9:5, where an account is demanded for the shedding of blood, human or animal, accenting the sanctity of life—non-human creatures inherently sustain their *nephesh.* This renders them as "pure souls," existing in a state of ontological integrity with their life force.

In his analysis of Job 12:7–10, Levison expands *nephesh* to encompass animals, emphasising a universal dependence on God's sustaining breath across all living beings.[12] Job 12:7–10 suggests that the Spirit's presence is discernible within the natural world, as the spirit of every creature, including animals, testifies to the Creator's ongoing activity:

> But ask the animals, and they will teach you;
> the birds of the air, and they will tell you;
> ask the plants of the earth, and they will teach you;
> and the fish of the sea will declare to you.
> Who among all these does not know
> That the hand of the Lord has done this?
> In his hand is the life *[nephesh]* of every living thing
> And the spirit *[ruach]* of every human being. (Job 12:7–10)

Here, *nephesh* signifies the essence of life, while *ruach* underscores the sustaining divine spirit. Together, they disclose an intimate kinship between all creatures and the Spirit who holds every living thing in being.

Ezekiel 37 presents a compelling theological interaction between *nephesh* and *ruach*, with the divine breath symbolising God's restorative power over life itself. Although this passage is traditionally read as a metaphor for Israel's national rebirth, both Macchia and Moltmann argue for a broader, more universal application.[13] They consider that the vision's invocation of *ruach* extends beyond human concerns to

[12] Levison, *Filled with the Spirit*, 17. See also Moltmann, *The Spirit of Life*, 46.

[13] See Moltmann, *The Spirit of Life*, 54–57., Macchia, *Jesus the Spirit Baptizer*, 126–27.

encompass the entirety of creation, thereby implicating all living beings—humans and animals alike—as recipients of God's sustaining Spirit. In this view, Ezekiel's prophecy gestures towards an eschatological renewal that transcends anthropocentric boundaries, highlighting a cosmic scope wherein the Spirit's life-giving force permeates all of animate existence.

In his analysis of Ezekiel, Michael Welker argues that the creative Spirit sustains fleshly life by imparting a share in the life-giving force that is common to all beings. The Spirit's work, therefore, is not abstracted from the frail and perishable nature of physical existence but engages directly with it. Welker suggests that the Spirit's renewing action in creation does not erase the inherent vulnerability of flesh; instead, renewal involves a transformation within the very fabric of fleshly life. The Spirit's creative work is thus intimately connected with the material world, effecting change and renewal without bypassing the reality of human and animal mortality.[14]

Flood Narrative

Transitioning from broader considerations of *ruach* and *nephesh*, the flood narrative offers a concrete illustration of the Spirit's covenantal involvement with creation. The flood narrative in Genesis (6:3, 17; 7:22; 8:1) not only reflects the creative power of the Spirit of God but also establishes a covenantal relationship that includes both humans and animals.

[14] Welker, *God the Spirit*, 164. Adding to this universal aspect of the sustaining Spirit, Psalm 104 provides a theological vision of creation's dependence on God's sustaining Spirit. In verse 29, the psalmist writes, "When you send forth your Spirit (*ruach*), they are created, and you renew the face of the ground," presenting *nephesh* as a concept that encompasses all living beings, not only humans. The term *ruach*, representing God's Spirit, conveys the divine breath that animates and renews creation. This suggests that animals, like humans, are sustained by the Spirit.

While my primary analysis centres on the creation account, the theological motifs of *nephesh* and *ruach* recur with significance in the New Testament, extending their implications to all living beings. In John 20:22, for instance, the risen Christ breathes the Holy Spirit onto his disciples, symbolising a transfer of divine life and empowerment reminiscent of the *ruach* that animates in Genesis. Likewise, Paul's statement in Acts 17:25—"God gives life and breath to everything"—uses inclusive language that implies God's life-sustaining breath encompasses all of creation, transcending anthropocentric boundaries. Together, these passages contribute to a broader theological vision where *nephesh* and *ruach* reflect a shared vitality infused by the Spirit across all creatures, resonating with the interwoven life force found throughout Scripture. This continuity suggests a theological framework in which animals, like humans, participate in the Spirit's animating and sustaining presence. See also Edwards, *Breath of Life*, 37.

Following the flood, God's covenant with Noah extends to "every living creature" (Gen 9:9–10),[15] marking a significant moment where animals are explicitly brought into a divine promise of protection, preservation, and an enduring commitment promised to all creaturely life. This covenant affirms that the Spirit's work is not limited to humanity but encompasses the entire animal world, as seen in Levison's recognition that animals, too, are sustained by God's Spirit and participate in God's presence.[16]

The Spirit's manifestation as a dove upon the waters after the flood symbolises not only peace but also the Spirit's ongoing commitment to preserving life in all its forms.[17] Just as *ruach* and *nephesh* reveal the Spirit as the animating breath that infuses every creature with life, the dove signifies the Spirit's active role in upholding creation's relational integrity. Through the preservation of both human and non-human life, the Spirit upholds the promise to sustain creation, ensuring that all creatures are encompassed within God's redemptive and life-giving work. This preservation reinforces the Spirit's role as a continuous, protective force within creation, honouring the relational integrity of all living beings

[15] Or, as Bauckham puts it, it is a covenant between God and "all flesh." Bauckham, *Living with Other Creatures*, 224.

[16] Levison, *Filled with the Spirit*, 17.

James David Audlin presents a speculative yet evocative interpretation of the divine name YHWH, connecting it to the act of breathing. Audlin suggests that the name YHWH, traditionally seen as unpronounceable, may be most authentically expressed through a gentle exhalation—a soft breath that aligns with the rhythm of inhaling and exhaling. Drawing from Jewish mystical traditions, he posits that each breath serves as a sacred link to God, embedding the divine name within the very act of living. For Audlin, this interpretation frames the Spirit (or *ruach*) as a life-giving presence, accessible to every creature through their own breath. While not an academically peer-reviewed source, this notion invites reflection on the theological implications of *ruach* as an animating force that imbues all creatures with a shared vitality. Audlin's insights, albeit from a non-academic lens, resonate with the biblical ideas of *ruach* and *nephesh*, both of which affirm a universal life essence that bridges humans and animals. If, as Audlin suggests, each breath could be seen as an utterance of the divine name, then all breathing creatures participate in a kind of unspoken, cosmic liturgy. This perspective broadens traditional understandings of *ruach* and *nephesh,* contributing to a theology where animal life, too, reflects a form of worship—a continuous praise that transcends verbal language and invites creatures of all kinds into communion with the divine. Audlin, "Breathing the Name of God: 'YHWH' and 'Elohim' in the Gospel of John," 1–3.

[17] For Ambrose of Milan, the dove's return not only symbolises the end of divine judgment but also reflects the Spirit's commitment to preserving life. Ambrose, *On the Mysteries and The Treatise on the Sacraments*, Book 1, Ch 3.

in the divine covenant and illustrating a cohesive vision of *ruach* as a life-sustaining, covenantal force that extends to animals.[18]

Theological Considerations

Turning now to the theological horizon, the biblical terms *ruach* and *nephesh* invite us into a vision of life where God's breath is not confined to humanity but courses through the whole of creation. Far from being reserved for the *imago Dei*, divine breath animates all living creatures with the same vitality and care. This reading resists anthropocentric interpretations of Scripture and affirms a pneumatology expansive enough to include the animal kingdom within God's sustaining presence. Animals, too, breathe the Spirit's breath; they participate in the same life-giving force that imparts meaning and purpose to humanity. Such a view reframes creation as a shared economy of divine life, one where creaturely existence is not tangential but integral to the unfolding work of the Spirit.

Origen is often associated with a dualistic outlook and an apparent indifference towards non-human animals, whom he dismissively refers to as "dumb beasts." Yet, his teachings do not entirely negate the value of these creatures. Origen acknowledges that all animals are Spirit-breathed beings, possessing a soul. However, he confines the Spirit's work in animals to the mere animating force that governs their life and movement.[19] This restricted view contrasts with Origen's broader emphasis on the human experience of the Spirit, where the rational soul—endowed with free will—can be transformed by the Spirit of God into something richer and fuller. However, restricting the Spirit to humanity (as Origen does) undercuts the biblical witness of *ruach* as the very breath

[18] In his exploration of the Spirit's nature, Mark Wallace presents the provocative idea of the Spirit as an avian life form, emphasising the Spirit's tangible, life-giving presence within creation. This perspective challenges traditional Western theological interpretations that have predominantly focused on the Spirit's role within the divine Trinity and human salvation, often neglecting the Spirit's cosmic function as the breath that sustains all life. While the Spirit has historically been understood as both the Spirit of God and the Spirit of creation, Christian theology has typically stressed its divine and salvific aspects, overlooking its intimate relationship with the non-human world. Wallace's approach highlights the Spirit's vital role within the natural order, suggesting that animals, as fellow participants in the Spirit-filled creation, share in a form of relationality and agency often downplayed in theological discourse. See Wallace, *Fragments of the Spirit*, 136–37.

[19] Origen maintains that the Spirit of God inspires transformation through the rational soul and free will, but only for humans. On *First Principles*, 154.

of life shared by all flesh. For my purposes, Levison's broader reading offers a more generative foundation: it allows us to speak of animals not as passive recipients of mere "vitality," but as active participants in the Spirit's creative and sustaining work.

Gregory of Nazianzus, by contrast, embraced a more expansive pneumatology.[20] Gregory saw the Holy Spirit as the agent of free movement, dynamic and unconfined. For him, the Spirit did not dwell solely in humanity but actively animated all creation. The Spirit's life-giving presence touched the angels, oversaw all things, and brought motion and vitality to all living beings. For Gregory, it was the Spirit that infused life into all things, that allowed all life to sing the praises of its Creator.[21] In this understanding, the Spirit became the vivifying force not just within human souls but in the very fabric of creation, bestowing life wherever it moved.[22]

Although Origen and Gregory of Nazianzus held differences in scope on the Spirit's role in non-human creatures, both agreed on the importance of the Spirit as the source of life and movement for animals.

Similarly, Irenaeus believed that the Spirit's influence was not limited to human bodies but extended across all creation. In Irenaeus' view, eschatological renewal through the Spirit was not confined to humanity but encompassed an ecological dimension.[23] All creation—human and non-human—was to be renewed, sustained, and blessed through relationships

[20] Bergmann, *Creation Set Free*, 261. Bergmann explains that the emphasis on the Spirit's permeation, indwelling and consummation of creation in Gregory of Nazianzus' cosmic trinitarian pneumatology lends to the idea of a liberated creation.

[21] Nazianzus, *St Gregory of Nazianzus: On God and Christ: The Five Theological Orations and Two Letters to Cledonius*, 55–56. Gregory of Nazianzus invites his audience to observe the behaviour and characteristics of different creatures, such as fish, birds, grasshoppers, swans, and peacocks, and to contemplate the intelligence and beauty displayed in their actions. Gregory prompts his listeners to question the source of their melody, their love of beauty, and their skills in work and art. He suggests that these qualities are not arbitrary but are evidence of a Creator who endowed these creatures with unique features and abilities.

[22] Bergmann, *Creation Set Free*, 166.

[23] Irenaeus envisions the restoration of all creation in the eschaton, where not only humans but also animals will participate in a renewed, harmonious order. He describes animals returning to their pre-Fall state, subject to humanity and feeding peacefully on the produce of the earth. This reflects Irenaeus' broader eschatological vision in which the natural world is transformed and violence is abolished, symbolised by the lion feeding on straw. While Irenaeus does not explicitly connect this to the work of the Holy Spirit in this

with the Spirit. For Irenaeus, the Spirit was an agent of healing and restoration, offering living water to restore bodies fractured by sin. Ultimately, the Spirit's grace poured into the earth to sustain and restore bodily life, reflecting the unity of God's presence in the world.

Denis Edwards's theology in *Breath of Life* presents a vision of the Spirit as the animating and sustaining force within all creation, a presence that infuses life, purpose, and divine connection into every creature.[24] Edwards moves past the well-worn path of anthropocentric readings and instead foregrounds a universal, relational Spirit. This Spirit, as God's breath and word, is not just present but woven into the sinews of life itself, binding humans and non-humans in a Spirit-soaked ecosystem that is as cohesive as it is diverse. This vision is rooted in biblical sources, which Edwards interprets as affirming the Spirit's intimate and constant participation in the dance of life with all creatures.

In his reading of Psalm 104, Edwards finds a compelling example of this pneumatological framework. Here, the Psalmist describes how God's Spirit enlivens and sustains all creatures, breathing vitality into every aspect of the natural world.[25] Edwards notes that the Spirit's presence is not a fleeting moment of divine interaction; rather, it is the constant source of renewal and sustenance. Every creature depends on this life-giving breath, and when God "takes away their breath, they die and return to the dust" (Ps. 104:29). This cyclical dynamic, where God's Spirit both gives and reclaims life, reinforces Edwards' view of creation as unified and sustained by divine presence, in which humans and animals alike participate in and benefit from the Spirit's nurturing care. This imagery reinforces the theological significance of ruach and nephesh as expressions of a Spirit-infused life that transcends human-centred interpretations.

passage, his theological framework, particularly in *Against Heresies* Book 5, chapters 32–33, links the Spirit to the renewal and sustenance of all creation. This supports the idea that the Spirit's role in redemption extends beyond humanity, encompassing all creatures in the renewal of creation James R. Payton, *Irenaeus on the Christian Faith: A Condensation of Against Heresies* (Eugene, OR: Wipf and Stock Publishers, 2011), http://ebookcentral.roquest.com/lib/harvest/detail.action?docID=5706733.

[24] Edwards explores this idea in detail, particularly in his chapter "Breathing Life into a Universe of Creatures" (*Breath of Life*, 33–49), where he examines the Spirit's role as the breath that enlivens and sustains all creation. Here, Edwards expands on the theological significance of *ruach* as the life-giving force that binds all creatures into a shared experience of divine presence.

[25] Edwards, *Breath of Life*, 35–36.

Further enriching his argument, Edwards draws on early Christian theology, notably the insights of the Patristics, to strengthen his argument.[26] Figures like Basil of Caesarea and Ambrose of Milan offer theological foundations for understanding the Spirit as Creator and Sustainer. Basil, for instance, describes the Spirit as the one who "fosters the diversity" of creation, nurturing each form of life in its unique identity yet holding all in unity.[27] Edwards uses Basil's ecological sensitivity as a historical anchor for his argument that the Spirit does not simply enliven but actively participates in a web of relationality among all living beings. For Edwards, *ruach* and *nephesh* are not merely metaphors for life but are expressions of God's active, sustaining relationship with creation.

Through these images, Edwards draws us into a vision of *ruach* and *nephesh* where animals rise as vivid, breathing presences within the Spirit's ceaseless work of creation. They are not merely granted life but threaded into a Spirit-soaked ecology—each creature bears its own cadence in the divine order. Edwards' engagement with biblical texts and early Christian thought invites a reimagination of traditional views, positioning the Spirit as the connective, transformative force that brings all of creation into a harmonious, life-sustaining communion. This theological vision is a call to see animals—and indeed, all creatures—as integral participants in God's creative and redemptive work, encouraging a broader, more reverent awareness of the Spirit's ever-present work in the world.

Edwards' pneumatological framework provides a foundational understanding of the Spirit's intimate and ongoing involvement with all of creation. Moving from this perspective, Michael Welker focuses on the relational aspect of the Spirit's work within creation. Welker sees the Spirit of God as the means through which God's power becomes knowable, a force revealed both "in and on" human beings and their fellow creatures. This revelation of divine power occurs *through* them and *for* them, embodying the Spirit's work in a way that unites all creatures in life-sustaining relationships.[28] In this view, the Spirit's presence is not confined to isolated moments of empowerment but rather pervades creation as a continuous, life-giving force. It is the Spirit that enables creatures to enter into "rich, fruitful, life-sustaining, fortifying, and protective

[26] Edwards, 39–43.

[27] Edwards, 41–42.

[28] Welker, *God the Spirit*, 2–4.

relations" with one another, echoing the creative power of God that animates all life. Through this lens, it can be argued that animals, too, have a means of knowing God. The Spirit's presence and activity make God's power knowable, not only to humans but to all creatures.

Building on this, Welker suggests that the Spirit's action within creation is integral to upholding the vitality of all living beings, as it fosters a complex ecosystem of mutually enriching relationships. However, as Welker observes, discerning the Spirit's work within these reciprocal relationships can be challenging. The Spirit's life-giving influence is often subtle and interwoven within natural and creaturely interactions, making it difficult to distinguish these divinely guided connections from ordinary relations. Nonetheless, through this life-giving force, the Spirit continually acts to sustain, protect, and fortify creation, illuminating a divine presence that upholds the interconnectedness of all life.

As Welker articulates the Spirit's sustaining relationships, Sigurd Bergmann expands on this interconnectedness, suggesting that the Spirit transcends natural limitations and is not bound by mechanistic laws. Where Welker focuses on the Spirit's sustaining power within relational ecosystems, Bergmann emphasises the Spirit's ability to move freely throughout creation, infusing it with life beyond spatial and formal constraints.[29] This idea of the Spirit's boundless movement complements Welker's emphasis on interdependence and sets the stage for a richer theological appreciation of divine immanence.

Jürgen Moltmann, drawing on a similar theme of divine immanence, deepens the theological perspective by exploring how God's love is expressed through the Spirit's active presence within creation. For Moltmann, the Spirit's engagement with all creatures signifies a divine love that not only gives life but remains intimately connected to the creation it sustains. Thus far, these theological insights prompt a consideration of not only the Spirit's sustaining role but also the relational nature of God's presence, moving from an abstract sustaining force to an embodied, loving connection within the cosmos.[30] Moltmann sees *ruach* as both the imminent and transcendent presence of God, describing it as a dynamic

[29] Bergmann, *Creation Set Free*, 170. This vision of the Spirit aligns with Edwards' concept of communion, suggesting that God's creative breath sustains creation moment by moment, forming a network of living relationships unified by divine love, Edwards, *Breath of Life*, 130–42.

[30] Moltmann, *The Spirit of Life*, 50.

force that supports and energises all living beings.[31] In Moltmann's interpretation, *ruach* is not merely an abstract energy but a tangible divine presence that manifests as both divine power and open space for life to flourish. This "space of freedom" created by *ruach* allows creatures to experience life in fullness, embodying God's sustaining power as well as an encompassing reality in which all creatures, including animals, thrive.

Like Moltmann, Wolfhart Pannenberg offers a generous pneumatological view that affirms non-human creatures are integral to God's good creation.[32] In his systematic theology, Pannenberg makes key reference to the Spirit's animating presence within non-human creatures. While his consideration may have humans primarily in view, Pannenberg upholds that the "life, movement and activity" of non-human creatures can be attributed to the Spirit of God.[33] For Pannenberg, the Spirit is the integral creative presence of God within creation, as the Spirit animates life and unites all creatures in fellowship with their Creator.[34] The beauty of Pannenberg's pneumatology is that he does not divorce the Spirit from the creaturely realm; rather, his account of the "continuous creation"attests to the giving and sustaining presence of the Spirit in all of created life.[35]

Each of these perspectives presents a rich picture of the Spirit's role in the cosmos. What these perspectives collectively suggest (and what I want to draw out) is that any pneumatology attentive to animals must not reduce them to peripheral elements of creation. Rather, animals belong within this Spirit-filled matrix of life, beauty, and purpose. My own project, therefore, takes up the Spirit's cosmic presence as described by these thinkers, but presses further: if creation is Spirit-drenched, then

[31] Moltmann, *God in Creation (The Gifford Lectures, 1984–1985)*, 10–11. See also Moltmann, *The Spirit of Life*, 40–42.

[32] This description of Pannenberg's pneumatology also appears in the literature review section of this book. The repetition here serves to reinforce the central themes in Pannenberg's work that are foundational to the subsequent analysis. See Chapter 1.

[33] Wolfhart Pannenberg, *Systematic Theology Vol 2*. Translated by Geoffrey W. Bromiley (London: T&T Clark, 2004), 76.

[34] Pannenberg, *Systematic Theology Vol 2*, 32.

[35] Pannenberg, *Systematic Theology Vol 2*, 76. Pannenberg's account of continuous creation refers to the 'preservation' of creation, where creation is not a one-time historical act but a continual work of God in the Spirit.

animals too must be seen as embraced within this divine economy, their lives mattering not instrumentally but intrinsically.

Yves Congar offers a fitting culmination to these ideas by extending the significance of *ruach* to embrace all of creation, including animals. Congar's work resonates with the previous scholars while challenging narrow interpretations of the Spirit. Congar argues that the Spirit's breath is not simply an impersonal life force but an active divine presence that connects all creatures, human and non-human alike, in communion with God. This interpretation of *ruach* as the embodiment of God's creative and sustaining work affirms animals as participants in the Spirit's life-giving presence.[36] By challenging a Stoic view of *pneuma* as merely a static cosmic force, Congar advocates for a theological understanding of the Spirit that upholds the dignity and divine relationship inherent in all creaturely life, further underscoring the Spirit's active role in maintaining the sacred value of animals.

Feminist theologian Elizabeth Johnson extends Congar's view by exploring *ruach* as an invisible yet powerful natural force akin to the movement of wind.[37] She connects the wind to wings, highlighting the Ancient Near Eastern symbol of the bird as representative of divine female power. In Christian theology, the dove becomes a symbol of the Holy Spirit, linking its imagery to pre-Christian traditions of divine feminine presence. Johnson emphasises the Spirit's maternal imagery, such as hovering over primordial chaos in Genesis, sheltering in protective wings, and empowering Jesus during his baptism.[38]

For Johnson, *ruach* also signifies the breath of life shared by humans and animals, sustaining creation and connecting spirit and matter in profound unity. Johnson highlights that the Spirit's presence transcends dualistic notions of divine separateness, instead working cooperatively within creation to sustain and empower life.[39] This relational dynamic underscores *ruach* as a vital, creative force that animates the entire community of creation.

[36] Brooks, "Breathing Forth the Word: Yves Congar's Articulation of the Activity of the Holy Spirit in the Life of Christ.," 201.

[37] Johnson, *Women, Earth, and Creator Spirit*, 45.

[38] Johnson, 46.

[39] Johnson, 46–47.

Johnson's view compliments Indigenous-Christian theologies that understand the Spirit's life-breath as the animating force that imbues all living beings with vitality and purpose.[40] This view, while resonating with the spiritual insights of ancient cultures, has been further developed by indigenous scholars working within Christian theological traditions.[41] These scholars articulate a theology that centres on the inherent sacredness of creation, bringing forth a deeply relational spirituality. Rooted in the land and the creatures that dwell upon it, this framework places non-human animals within the same spiritual web of life that sustains human existence, a concept interwoven with the Spirit's omnipresence.

Randy Woodley conveys these themes by drawing attention to the Social Trinity—where the Spirit's relationality extends beyond humanity, encompassing all of creation in a harmonious bond. For indigenous Christians, the land is not merely a resource but a living, sacred space where the Creator's presence can be encountered. Theology, in this context, begins with the earth and its creatures, both human and non-human,

[40] In exploring the pneumatological theme of the breath of life across various cultures, it is valuable to acknowledge the insights of ancient and Indigenous belief systems, such as those of the Mesoamerican Mexica, which offer complementary understandings of life's animating force, much like Christian reflections on the Spirit. While the Mesoamerican peoples, commonly referred to as the Aztecs or Mexica, are not typically associated with contemporary theological discussions, their cosmology offers interesting insights into the essence of life that align with broader theological reflections on the Spirit and the breath of life. The Mexica understood life as fundamentally animated by the power of breath, a concept that resonates with the Christian theological notion of the Spirit as the giver of life. For the Mexica, the breath (known as *tonalli*) was the vital force that imbued all living beings—human and non-human—with life, emphasising that respiration was the defining characteristic of any living creature. Moreover, the Mexica had a sophisticated understanding of the relationship between breath, identity, and the afterlife. They believed that the breath was not only essential for life but also intimately tied to one's soul and fate after death. In this way, their beliefs about the animating power of breath extend beyond mere physical function, touching on deeper metaphysical questions about human and non-human life, identity, and existence after death. These reflections complement the ancient and indigenous Christian and non-Christian belief systems that similarly recognise the sacredness of breath as a life-giving force, further enriching theological discourse on the Spirit's role in all creation. Furst, *The Natural History of the Soul in Ancient Mexico*, 156.

[41] I recognise that my focus on a specific subset of Indigenous theologies does not capture the full breadth of Indigenous thought. While my framework emerges from a Christian systematic context, I regard Indigenous pneumatologies as vital dialogue partners that reorient us toward an ecological and creaturely belonging often lost in Western accounts.

who share in the Creator's life-giving breath. Woodley emphasises that within Indigenous worldviews, all creatures are alive, and panentheism—the belief that God is in all things yet distinct from them—reflects this interconnectedness.[42] This perspective places animals in a familial relationship with humans, as co-inhabitants of a world charged with the Creator's Spirit.[43] In this vision, God is not distant, not removed from the diversity and complexity of the world, but actively involved in it. Indigenous pneumatologies encourage viewing the world through this lens, to recognise that the Spirit's breath flows through every living being, binding creation together in a shared, sacred existence. Animals, in this framework, are not lesser beings but participants in the divine economy, sustained by the same Spirit that breathes life into all creation.

Garry Worete Deverell echoes Woodley and reconfigures the concept of the Spirit through an Australian Indigenous theological lens, situating the Spirit deeply within *country*—a term encompassing all aspects of the natural world, including animals, plants, waterways, and landscapes. In contrast to the hierarchical, human-centred view often found in Western theology, which emphasises the Spirit's procession from the Father and the Son, Deverell portrays the Spirit as a centrifugal force that binds all beings—human and non-human—through patterns of reciprocity, kinship, and care.[44]

For Deverell, the Spirit is not an abstract or distant entity but the very life force that animates and connects all forms of life within the material world. Animals are not external to the Spirit's work but are fully integrated into the relational web of creation, co-existing with humans in a sacred, interconnected world. The Spirit's role is to sustain these relationships between living beings, reflecting an Indigenous cosmology in which

[42] Woodley, *Indigenous Theology and the Western Worldview*, 77. Panentheism, then, becomes a key theological lens through which to understand the relationship between God and creation. It is a concept that refuses to separate the Creator from the created world, insisting instead that God is both transcendent and immanent. The analogy of the soul and body is often used to capture this idea: just as the soul permeates every aspect of the body, sustaining its life while remaining distinct from it, so too does God inhabit every part of creation while transcending it. This image conveys a sense of intimacy between God and the material world, a closeness that imbues even the smallest creatures with sacred significance.

[43] Kidwell, Noley, and Tinker, *A Native American Theology*, 39;127.

[44] Deverell, *Contemplating Country*, 21.

animals are kin and share a reciprocal relationship with both humans and the land.[45]

Deverell further suggests that the Spirit may be understood as the *voice of country*, where animals, alongside elements of the landscape such as mountains and streams, communicate the wisdom and guidance of the creator ancestors. In this framework, animals are not passive objects but active participants in creation, animated by the Spirit and essential to the fabric of life. This perspective aligns with Indigenous cosmologies that grant animals agency and purpose, positioning them within a web of interdependent relationships with humans, the land, and the Spirit.

Likewise, Deverell's imagery of the Spirit as the *aliveness of country*—manifested through breath, water, and fire—further suggests that animals are animated by the same life-giving force as humans. This understanding rejects the dichotomy between the spiritual and material realms, affirming that animals are as sacred as human beings. The Spirit is present in the very life force of animals, embodying the vitality of creation itself. Relationships between humans and animals are thus not merely functional but deeply spiritual, shaped by the Spirit's binding force of kinship and care.[46]

From this standpoint, animals are not to be dominated or used by humans; instead, they are kin, animated by the same Spirit that gives life to all creation. The presence of the Spirit in animals underscores a broader relational framework in Indigenous theology, in which humans and animals share a sacred bond, each contributing to the flourishing and balance of creation. Deverell's perspective calls for a recognition of the Spirit as a dynamic, life-sustaining force that flows through all living beings, fostering a deep responsibility towards animals as integral members of the spiritual and material community of creation.

Such a position challenges anthropocentric theology, calling for a rethinking of the place of animals within Christian thought. If animals, too, are alive with the Spirit's breath, then their value extends far beyond

[45] Deverell, 22.

[46] Deverell's view of the Spirit as a life force that connects all living beings echoes the L'nuk understanding of animals as beings with spirit, alongside elements like rocks, water, and fire. In L'nuk culture, animals are seen as equal and sacred, participating in a shared life force. This is reflected in the practice of *netukulimk*, which guides respectful harvesting, ensuring that animals are taken only as needed, in recognition of their spiritual value. The eagle, in particular, holds deep spiritual significance, often seen as carrying prayers to the Creator, demonstrating the sacred relationship between humans and animals. Robinson, "L'nuwey Views of Animal Personhood and Their Implications," 26–27.

their utility to humans. They are living creatures, spiritual beings whose existence is bound up in the Spirit's sustaining presence, whereby virtue of their ontology, possess an inherent and unbroken connection with their Creator. Indigenous pneumatology offers a compelling vision of a world where all of creation, human and non-human alike, participates in the life of the Spirit—a vision that calls for a deeper respect and reverence for animals.

Animal Ontology: Sentience and Spiritual Potential

Building upon the foundational concepts of *ruach* and *nephesh*, the discussion now turns to their implications for understanding animal sentience and spiritual potential.[47] If animals share in the Spirit's animating breath, then their capacity for subjective experience and relational engagement with God deserves deeper consideration.

Animal sentience highlights the notion that animals, as creatures imbued with autonomy and integrity, are co-companions to humanity in the broader narrative of creation, reconciliation, and redemption. Sentience refers not merely to biological responsiveness but to an animal's capacity to experience a range of emotions—joy, pain, contentment—that signal an inner life and relational awareness akin to that of humans.[48] While human moral behaviour is complex, animals exhibit behaviours that resonate with human-like expressions of care, play, and even discipline, underscoring their role as subjects rather than objects within creation.[49]

Acknowledging the distinctions between humans and animals need not obscure the similarities in our capacities for shared experience and emotional depth. This theological position on sentience challenges reductionist views by recognising animals as creatures capable of relational depth and participation in the Spirit-filled life of creation.

[47] A version of this section has been published in Rizzo, D. "Towards a More Holistic Sustainability: Animal Sentience and Spiritual Potential", in Luetz, J.M., Nunn, P., & Filho, W.L., *Handbook of Traditional Spirituality and Sustainability* (pp.). Springer.*Forthcoming.*

[48] Linzey, *Developing Animal Theology*, 1–2.

[49] See Marc Berkoff, "The Evolution of Animal Play, Emotions, and Social Morality: On Science, Theology, Spirituality, Personhood and Love," *Zygon: Journal of Religion & Science* 36, no. 4 (December 2001) for a comprehensive exploration of these themes.

For this section, I employ an interdisciplinary methodology that integrates insights from both scientific research and theological reflection. By drawing on empirical studies of animal behaviour, cognition, and sentience, I seek to ground the discussion in a robust understanding of animals as complex, emotionally responsive beings. This scientific foundation provides a credible basis for exploring concepts like emotional depth, relational behaviours, and communicative complexities in animals. These findings serve as a springboard for theological reflection. I will interpret the empirical evidence through a pneumatological and eschatological lens, informed by *ruach* and *nephesh*. Theologically, I aim to understand animal sentience not merely as biological responsiveness but as a potential mode of participating in the Spirit's relational and life-sustaining presence. This approach enables a holistic examination that attends to both the observable realities of animal life and their theological implications, moving beyond anthropocentric limitations to propose a framework that honours animals as integral members of creation with inherent spiritual potential.

Animal Sentience

Recognising animals as sentient beings with a subjective inner life is essential for defining and understanding their place in the world. This perspective challenges traditional views that have often reduced animals to mere resources or instruments for human use.[50] There is an emerging consensus among scientific researchers that animals are sentient beings with a subjective inner life. Complex experiences are not limited to mammals and birds, as updates to the Cambridge Declaration on Consciousness suggest that a wider range of animals, including some invertebrates, may exhibit conscious states.[51] While I draw a strict sentience line for this book, I recognise that the scope of sentience

[50] To reference historical perspectives on the instrumentalisation of animals for human use, consider the following resources that explore this shift: Murray, "The Underdog in History: Serfdom, Slavery and Specis in the Creation and Development of Capitalism"; Shannon, "Invisible Parts: Animals and the Renaissance Anatomies of Human Exceptionalism." These works provide insight into the historical development of human-centred views that reduced animals to resources or instruments, tracing the roots of these perspectives in capitalist and Renaissance contexts.

[51] Falk, "Insects and Other Animals Have Consciousness, Experts Declare."

remains an ongoing area of scientific exploration.[52] Sentience encompasses an animal's capacity to experience both positive and negative emotions, including pain, joy, contentment, and various forms of mental suffering such as fear, shock, terror, anticipation, anxiety, stress, foreboding, or distress.[53]

Research conducted by Murata et al. offers a striking glimpse into animal sentience through the lens of dogs' emotional responses and nonverbal communication. Their study traced the subtle connection between emotional arousal and tear production, revealing that dogs, like humans, secrete more tears during moments charged with feeling—such as joyful reunions with their owners.[54] This physiological response, long associated with human emotion, suggests that canine experience is textured with joy, attachment, and the ache of reunion. The role of oxytocin (the bonding hormone) proved decisive here: its influence on tear secretion highlights how deeply emotional connection is etched into the biological fabric of the dog–human bond.

What makes these findings even more compelling is that they do not stand alone. A parallel study on rats uncovered a similar coupling of emotion and physiology, this time in the form of ultrasonic vocalisations (tiny cries inaudible to the human ear) that were likewise driven by oxytocin.[55] These vocal signals function as social cues, allowing rats to express positive emotional states and strengthen affiliative ties. Taken together, the studies extend beyond the familiar terrain of domesticated animals, pointing towards a wider mammalian repertoire of emotionally mediated communication.

In both dogs and rats, oxytocin emerges not only as a biochemical marker but as a bridge: a mechanism that converts inner feeling into outward expression, whether through tears in a dog's eyes or ultrasonic chirps echoing in a rat's enclosure. Such findings add weight to the claim that animals are not merely reactors to stimuli but sentient beings,

[52] See Chapter 1.

[53] Bekoff, *The Emotional Lives of Animals.*

[54] Murata et al., "Increase of Tear Volume in Dogs after Reunion with Owners Is Mediated by Oxytocin."

[55] Gloveli, Natalie, Jean Simonnet, Wei Tang, Miguel Concha-Miranda, Eduard Maier, Anton Dvorzhak, Dietmar Schmitz, and Michael Brecht. "Play and Tickling Responses Map to the Lateral Columns of the Rat Periaqueductal Gray." *Neuron* 111, no. 19 (October 4, 2023): 3041–3052.e7. https://doi.org/10.1016/j.neuron.2023.06.018.

equipped with emotional depth and social tools that tether them to others in bonds of care, attachment, and joy.

Primatologist and ethologist Frans de Waal explores the connection between evolved behaviour, especially in non-human primates, and the concept of normativity within the realm of morality.[56] He posits that animals, including non-human primates, exhibit behaviours indicative of their commitment to ideals or standards in their social interactions, akin to the way humans follow moral values. They actively seek to preserve social harmony through actions like reconciliation after conflicts, protesting against unfair distributions, and intervening to quell disputes among peers. His research suggests that these normative behaviours are underpinned by emotional values, akin to human morality's emotional foundations. Additionally, he discusses impulse control, goal-oriented behaviour, and the role of punishment in shaping social norms among animals, demonstrating an understanding of and adherence to social norms and ideals, akin to human moral behaviour.[57]

Situating animals as sentient implies that non-human creatures are capable of experiencing a wide range of emotions and displaying behaviours akin to human conduct. Even though these experiences may be different, such acknowledgement does not diminish animals to engage in a form of spirituality. Recognising sentience serves as a foundational step in this chapter. By exploring the concept of animal spirituality as part of the ontological essence of non-human creatures, it highlights the commonality of their inner lives and paves the way for a deeper understanding of their spiritual capacity.

Animals as Communicators

The scientific study of animal communication has gained momentum in recent years, with increasing evidence pointing to the existence of intentional communication among various animal species.[58] While animal communication does not possess the formal properties of human language, some experts suggest that it exhibits elements that could serve

[56] De Waal, "Natural Normativity."

[57] De Waal, "Natural Normativity," 200.

[58] Benz-Schwarzburg, *Cognitive Kin.*

as precursors to human language. Kristen Andrews argues that understanding animal communication requires considering the unique social, biological, and ecological contexts of different species.[59] There is a lack of consensus regarding the nature of the conveyed messages in animal communication.[60]

There is a growing body of research in the fields of animal behaviour and communication that suggests that many species engage in intentional communication.[61] Although there is debate about the precise nature and scope of animal communication, a number of leading scholars and researchers endorse this perspective. Frans de Waal has written extensively on animal behaviour and communication and argues that many species engage in intentional communication that is both sophisticated and nuanced.[62] He has studied the communication of chimpanzees, bonobos, and other primates, and has argued that their vocalisations and gestures can convey complex meanings and emotions.[63]

Irene Pepperberg, a comparative psychologist, has conducted groundbreaking research on the communication abilities of African grey parrots. Her work has shown that these birds are capable of using words and symbols to communicate with humans in a way that suggests intentional communication.[64] Biologist and ethologist Marc Bekoff has argued that many species, including wolves, coyotes, and elephants, engage in intentional communication through a variety of signals and behaviours. He has also written extensively on the emotional lives of animals and has argued that their communication is often linked to their social and emotional

[59] This idea is the thrust of Kristin Andrews, *The Animal Mind: An Introduction to the Philosophy of Animal Cognition* (London: Routledge, 2020).

[60] Andrews, 157.

[61] See Jardat, Liehrmann, and et al., "Horses Discriminate between Human Facial and Vocal Expressions of Sadness and Joy"; Pardo et al., "African Elephants"; Pokharel, Sharma, and Sukumar, "Viewing the Rare through Public Lenses."

[62] See De Waal, *Good Natured: The Origins of Right and Wrong in Humans and Other Animals*; De Waal and Andrews, "The Question of Animal Emotions"; De Waal and Ferrari, "Towards a Bottom-up Perspective on Animal and Human Cognition."

[63] De Waal, "Natural Normativity." 200.

[64] Pepperberg et al., "Logical Reasoning by a Grey Parrot?".

relationships with others.[65] While there is no indication that animals possess language with the same formal properties as human language, some ethologists suggest that animals may exhibit elements that could serve as precursors to human language.[66] These precursors include facial expressions, body posture, eye contact, vocalisations, gestures, touch, and mutual gaze.[67]

Although scholars generally accept that intentional communication exists among certain animal species, there is a lack of consensus regarding the nature of the conveyed messages. Some of these signals serve to capture attention, while others convey the emotional state of the sender. A variety of communicative signals, such as contact calls, food calls, alarm calls, whistles, displays of body parts, dances, gestures, vocalisations, touches, and eye contact, have been identified to gain attention. Some signals are used referentially, while others are imperative in nature.

[65] Marc Bekoff, "The Evolution of Animal Play, Emotions, and Social Morality: On Science, Theology, Spirituality, Personhood, and Love," *Zygon: Journal of Religion & Science* 36, no. 4 (December 2001): 615, https://doi.org/10.1111/0591-2385.00388.

[66] Kristin Andrews explores this perspective in detail. See Andrews, *The Animal Mind*.

[67] In her chapter on animal communication, Andrews demonstrates that the study of animal cognition necessitates an understanding of the unique social, biological, and ecological contexts of different species. Andrews summarises animal communication in the forms of vocalisations, signals, and language. In studying natural animal behaviour to detect evidence of intentional communication, Andrews considers three key factors: content, function, and intent. The content of the message, or what the signal conveys, is examined first. Next, the function of the message is evaluated in terms of the sender's desired outcome. Finally, researchers assess whether the signal is voluntary and under the control of the sender since intentional communication requires the ability to control the message. For example, elephants utilise sound communication that can travel over long distances, while electric fish rely on electric charges in water, and dolphins use echolocation, using clicks instead of eyesight to gain information about their surroundings and other dolphins. Andrews, 136–71. These communication cues are specific to each species and do not rely primarily on vision. Similarly, humans, who do not possess the ability to hear low-frequency sounds or sense electricity or magnetism, should not expect these modalities to play a role in human communication. While humans may not currently rely on magnetoreception or other non-visual modalities for communication, recent studies suggest we might retain some neural sensitivity to geomagnetic fields, indicating the possibility of an ancestral sensory system that is no longer consciously utilised. See Wang et al., "Transduction of the Geomagnetic Field as Evidenced from Alpha-Band Activity in the Human Brain."

Just as humans can convey emotions through the tone of their speech, certain animal signals also appear to reflect the sender's feelings.[68]

Human language is distinguished by its grammatical structure, which enables the creation of new meanings from familiar symbols. It is possible that animal communication systems, too, are more complex than they appear. Current evidence suggests that most animal communication systems may lack the flexibility of human language, though ongoing research continues to reveal new insights into the potential nuances of animal communication. Andrews examines whether gestural communication can provide syntax and productivity in communication and suggests that pantomime may have been the means by which humans first developed language.[69] While there is ongoing debate about the precise definition of natural language, research has refined the criteria used to evaluate linguistic capabilities in non-human animals. Benz-Schwarzburg proposes a minimal definition of natural language that includes symbolic systems exhibiting compositionality, systematicity, and productivity. For instance, attempts to teach language to animals, such as the bonobos Kanzi and Washoe, have demonstrated that while these apes can master basic symbols and some grammatical rules, they generally do not achieve a full grasp of complex grammatical structures or propositional knowledge.[70] Thus, the difference between human language and ape communication systems may be a matter of degree rather than a categorical distinction.

The study of animal communication confirms that intentional communication exists among many animal species. However, it differs significantly from human language, as the grammatical structure that allows for the creation of new meanings from familiar symbols is unique to human language. The conceptual framework and linguistic labelling that differentiate various species based on their physical, mental, and social characteristics may be useful for reference purposes. Yet these do not provide any insight into the subjective experiences of these living beings. These categories shape the way humans perceive and interact with other

[68] For a deeper exploration of animal communication and its emotional and referential dimensions, see Bekoff, *The Emotional Lives of Animals*; Pardo et al., "African Elephants."

[69] Andrews, *The Animal Mind*, 136–170.

[70] Benz-Schwarzburg, *Cognitive Kin*, 38–39.

species to such an extent that individual animals are often reduced to mere abstract concepts.

The use of animals in pain research and laboratory experiments to investigate cognitive processes traditionally associated with humans—such as attention, perception, associative and serial learning, working memory, spatial cognition, numerical processing, concept formation, and problem-solving—raises significant ethical and philosophical questions. While these studies challenge the notion that animals lack psychological traits comparable to those of humans, they also underscore the need for a nuanced understanding of both the scientific merits and moral implications of such research. This perspective invites a critical examination of the ethical frameworks guiding animal research, especially when compared to the moral considerations of conducting experiments on humans (which is beyond the scope of this essay). However, such practices demonstrate that animals exhibit complex mental processes and emotional experiences that closely mirror those of humans.[71] For example, for over 50 years, it has been known that mice emit ultrasonic vocalisations (USVs), high-pitched sounds inaudible to humans, which constitute a form of singing.[72] These vocalisations occur in various contexts, such as when a mouse pup calls for its mother, and become more complex as mice mature. Most female mice prefer spending time near speakers that play complex tunes. The varied reactions of the females to different songs further support the conclusion that these calls convey distinct meanings. Researchers are still decoding these songs and exploring how they vary across different social situations. Recent findings suggest that while mice have a more limited ability to modify their songs compared to songbirds, they nonetheless provide valuable insights into vocal communication and may be useful in studying communication disorders.[73]

Current research on animal cognition and emotion confronts the traditional view that denies humanity's connection to animals.[74] This

[71] Palmer et al., "What Do Scientists Mean When They Talk About Research Animals 'Volunteering'? In: Society & Animals - Ahead of Print."

[72] Chabout et al., "Male Mice Song Syntax Depends on Social Contexts and Influences Female Preferences."

[73] Chabout et al.

[74] Celia Deane-Drummond, *The Wisdom of the Liminal: Evolution and Other Animals in Human Becoming* (Grand Rapids, MI: Eerdmans, 2014) 51.; Rebekah Humphreys, *Animals, Ethics and Language: The Philosophy of Meaningful Communication in the Lives of*

denial, when taken to extremes, can lead to a form of human chauvinism that justifies ideologies of domination, subsequently resulting in slavery, genocide, the exploitation of women, and the mistreatment of animals.[75]

In acknowledging the sentience of animals and their capacity for subjective experiences, I move beyond abstract categorisations and recognise their potential for spiritual experience. This consideration not only aligns with theological views about the interdependence of all creation but also challenges harmful ideologies that perpetuate domination and exploitation. Embracing the emotional and cognitive complexity of animals fosters a more compassionate and ethical approach to our interactions with other species, thereby promoting a worldview that values and respects the subjective experiences of all living beings.

While animal communication is often unintelligible to humans, animals use vocalisations, signals, and gestures to convey messages to one another. The nature and complexity of these communications are not fully understood, and there is ongoing debate about how they compare to human communication. This raises intriguing questions about the possibility of animal spirituality and whether animals might possess forms of communication that extend beyond or differ from human language. As animal vocalisations, signals, and gestures are forms of communication that are unintelligible to humans, one might question whether these animal expressions are comprehensible to the Spirit. By exploring these aspects, we can better understand the potential for shared or distinct modes of expression and interaction across species. In the subsequent sections, I will consider the concept of animal spirituality and consider behaviours that may offer insights into how animals engage in practices that transcend mere communication.

Animals and the Expression of Spirituality

Spirituality has long been associated with human consciousness and reasoning. Yet recent studies suggest that animals, too, may possess spiritual experiences that are different from those of humans. Defining

Animals, The Palgrave Macmillan Animal Ethics Series (Switzerland: Palgrave Macmillan, 2023), 34–35.

[75] Kemmerer, "Theorizing 'Others.'"

animal spirituality solely through human-like characteristics such as self-transcendence and a search for meaning may therefore be unduly limiting. I also take into consideration that the argument for animal spirituality is predominantly based on analogies drawn from humans to animals, rather than the other way around. Here, I consider the potential for animal spirituality alongside the different definitions and understandings of this phenomenon. Defining animal spirituality solely through characteristics such as self-transcendence, identity quest, or a search for meaning presumes that reflective consciousness and human reasoning are primarily spiritual experiences. Human spirituality, which encompasses subjective experiences of value fulfilment, varies considerably across individuals and cultures. It is therefore reasonable to anticipate similar diversity in the expressions and manifestations of animal spirituality.

Animal spirituality may involve subcortical, affective, extra-linguistic, relational, organismic, and phenomenal experiences that do not require conversion to meaning through interpretive cognitive functions.[76] Unlike humans, animals may not reflect on their identity or question the meaning of their existence, as their identity and the meaning of their lives may be self-evident and intuitively comprehended.[77] As our understanding of animal cognition and behaviour continues to develop, it is likely that we will gain a deeper appreciation for the complex and nuanced lives of non-human creatures.

One definition of spiritual experience refers to any experience that promotes an organism's flourishing, increases the meaning or purpose of its life, and enhances the quality of the creature's existence.[78] Brooks Pribac defines spirituality as a phenomenon originating in the body and the ancient brain regions, such as the limbic system and brainstem, which are shared by humans and other animals.[79] These brain regions are involved in fundamental emotional and physiological processes that underlie spiritual experiences. This phenomenon is characterised by a natural inclination of a non-reflective, experiential consciousness that is

[76] Brooks Pribac, "Spiritual Animal."

[77] Paul Cunningham provides a comprehensive examination of animal spirituality across two articles, addressing key issues such as conceptual challenges, methodological considerations, and the question of animal consciousness. See Cunningham, "The Case for Animal Spirituality-Part 1.", and Cunningham, "The Case for Animal Spirituality-Part 2."

[78] Clough, *On Animals, Theological Ethics*, 2:23–25.

[79] Brooks Pribac, "Spiritual Animal." 354.

intrinsically relational and can be understood best in terms of affective vitality. If spirituality is both defined and scaffolded in this manner, then it can be concluded that animals do indeed possess the potential for spiritual experiences.

Recent interdisciplinary research trends indicate a shift in perspective that may lead to novel approaches for conceptualising and investigating proto-religious states in animals, as well as non-linguistic, non-conceptual forms of spirituality.

In 2005, Jane Goodall, a renowned English primatologist and conservationist, documented her observations on what she characterised as "sacred rituals" or expressions of spirituality in chimpanzees. Based on years of research, Goodall identified foot thumping, rock throwing, vigorous jumping, hooting vocalisations, and subsequent quiet sitting and staring behaviour in the presence of natural wonders such as waterfalls and rain as a "dance" accompanied by non-ordinary emotions similar to wonder and awe.[80] For Goodall, behaviours and emotions could be considered potential precursors to religious rituals and the origin of spirituality in human primates, leading to animistic and pagan worship of the elements. These observations, according to Goodall, demonstrate the possibility of proto-religious states in animals and suggest the existence of a non-linguistic, non-conceptual form of spirituality.

Mark Bekoff has observed foxes engaging in "ritualistic" behaviour around the bodies of dead foxes. He reports that live foxes will approach the body of the dead fox, sniff it, and then begin to bury it with leaves, dirt, and other materials. The live foxes will then sit near the burial site, sometimes for hours, and seem to mourn the loss of their companion.[81] Bekoff's observations suggest that the fox funeral is a complex behaviour that reflects the social nature of foxes. It is believed that the dead fox's burial protects the body from predators and prevents the spread of disease. Yet the fact that the live foxes stay near the burial site and appear to mourn the dead suggests that there may be an emotional dimension to this behaviour as well. For Bekoff, foxes are not the only animals to engage in what could be considered mourning or funeral

[80] Jane Goodall, "Do Chimpanzees Have Souls?: Possible Precursors of Religious Behavior in Animals," in *Spiritual Information: 100 Perspectives on Science and Religion*, ed. Charles L. Harper and Charles L. Harper Jnr (West Conshohocken: Templeton Press, 2005), 275–78.

[81] Bekoff, "The Evolution of Animal Play, Emotions, and Social Morality," 9.

rituals. Elephants, chimpanzees, and several other species have also been observed engaging in similar behaviours around their dead. These observations suggest that animals may have a greater capacity for emotion and social behaviour than previously thought.[82]

Although the extent to which animals exhibit spiritual or transcendent experiences is unknown, emerging research suggests that there may be more to animal consciousness and behaviour than traditionally recognised. The existence of structural and functional variations in animals indicates that we cannot determine the full nature and degree of differences that might impede the spiritual experiences of species which have evolved differently. This at least suggests that any claim of complete similarity between animal and human spirituality is likely to be unfounded. It is possible that the unique feature of animal spirituality, namely the ability to have a non-reflective, non-verbal, and conscious experience of fulfilling values, may not require a human brain or body.[83] The observations provided by Goodall and Bekoff suggest that animals exhibit complex behaviours and emotions which were previously thought to be unique to humans. Further research is needed to better understand the nature of animal spirituality and its implications for our understanding of consciousness and the natural world. Animals could potentially engage in forms of spirituality that are not conceptual, questioning the traditional view that spirituality is a uniquely human characteristic.

Biblical Perspectives on Animal Ontology

Sentience, when framed theologically, opens towards the possibility that animals possess a spiritual dimension. For Wolfhart Pannenberg, the Spirit constitutes God's creative and unifying presence within the world, animating life and drawing all creatures, human and non-human alike, into fellowship with God.[84] For Pannenberg, the Holy Spirit cannot be divorced from the creaturely realm, as the Spirit sustains all created life.

[82] Bekoff, *The Emotional Lives of Animals.*, 58. Bekoff highlights the growing scientific consensus that humans and animals share comparable chemical and physiological systems. This is evident in the frequent use of animals for developing and testing treatments for human mental health conditions, with studies indicating that mice, for example, can effectively model traits like sadness and introversion.

[83] Merker, "Consciousness without a Cerebral Cortex."

[84] Pannenberg, *Systematic Theology Vol 2.*

This theological perspective aligns with biblical references that point to animals participating in divine praise and lament. For instance, Psalm 148 calls upon all creation, including animals, to praise God. This passage suggests that animals can glorify God in ways unique to their nature, without requiring human language or reflective consciousness.[85] Furthermore, prophetic writings, such as those in Isaiah and Hosea, depict animals as capable of mourning and responding to God. This highlights their spiritual capacity and the relational depth they share with human creatures.

The biblical writings present a nuanced understanding of praise extending beyond human worshippers. The Psalms and prophetic writings call upon all creation, including animals, to offer praise and reverence to God (Ps 148, 145:21; Is 66:23; Rev 5:13). While some commentators limit these references to humans, the phrasing suggests that all living creatures are included in the call to worship God.[86]

Terence Fretheim explores the theme of creaturely praise in the Psalms where praise is ascribed not only to human worshippers but to the entire created world. Psalm 148 illustrates this motif, encompassing humans, angels, animals, plants, and inanimate objects, all created by God.[87] The text highlights a diverse spectrum of non-human creatures—from sea dwellers to wild animals, cattle, creeping things, and birds—depicting the all-encompassing praise of creation. Fretheim considers "the listing

[85] Atkins explores the role of animals in praising God as depicted in the Old Testament. He argues that animals are active participants in worship and that their praise of God is often portrayed as spontaneous and unselfconscious. Atkins, "Praise by Animals."

[86] For example, Psalm 150:6 is a declaration that links human praise with that of animals, stating, "Let everything that has breath praise the Lord." For Fretheim, biblical evidence implies that non-human creatures "have a certain inwardness or interiority such that more than external relationships are possible." Fretheim, *God and World in the Old Testament*, 255. See Fretheim's chapter dedicated to an exploration of nonhuman worship: Fretheim, 249–68. Atkins specifically argues that Psalm 150:6 should not be read as encompassing all of creation, as inanimate objects like rocks lack any connection to *neshamah* (breath). Instead, he suggests the verse may be limited to heavenly beings, humans, and animals. Atkins, "Praise by Animals," 506.

[87] Fretheim, *God and World in the Old Testament*, 257. This depiction is not simply a figure of speech but a revelation that animals are companions in the divine worship of God, Johnson, "Animals' Praise of God," 270.

together of the creatures" as suggesting both individuality and complementarity in praising.[88] Each animal praises according to its intrinsic capacity and fitness, with varying degrees of complexity.

Although animals may not express their praise in the same manner as humans, they possess a spontaneity and consistency that humans sometimes lack. Animals glorify God in their unique ways, independent of human language. For Fretheim, non-human animals have a voice and a role in praising God. Similarly, Johnson contends that animals praise God by existing as they were intended. While animals may not communicate in human language, their voices resound throughout the earth. This depiction, according to Johnson, reveals that animals are companions in the worship of God.[89] The psalmists discerned in the roar of lions, the flight of birds, and the mysterious movements of sea creatures expressions of reverence for a God who is saviour and judge of both humanity and the animal kingdom.

Bauckham endorses this position by affirming that every human and non-human being glorifies God by merely existing and carrying out its divinely assigned duties in God's creation.[90] Bauckham identifies an antiphonal praise pattern in various biblical texts, such as Psalm 148, where alternating lines of praise are offered between different groups of worshippers. He posits that this pattern reflects a communal and participatory form of worship, where human and non-human creatures actively engage in praising God together.[91] Bauckham suggests that this form of worship has implications for contemporary Christian worship practices, fostering a sense of community among all living creatures and shifting the focus of worship away from individualistic and anthropocentric expressions towards a communal, creaturely expression. This reading suggests that animals are partners in the cosmic worship of God. They are fellow

[88] Fretheim, "Nature's Praise of God in the Psalms," 23.

[89] Johnson, "Animals' Praise of God," 269.

[90] Bauckham, "Joining Creation's Praise of God," 47.

[91] According to Bauckham, antiphonal praise is a type of worship in which different groups or individuals alternate in offering praise to God. In this form of worship, there is a call-and-response pattern between the worshipers. Bauckham, *Living with Other Creatures*. Additionally, antiphonal praise can help to shift the focus of worship away from individualistic and anthropocentric expressions of praise towards a communal, creaturely expression of worship.

creatures with their own relationship to God, and they respond to the Creator's care in their lives.

The prophet Isaiah and the Psalms share a common theme of universal praise by "all flesh." Isaiah 66:23 proclaims that "from new moon to new moon, and from Sabbath to Sabbath, all flesh shall come to worship before me, says the Lord," while Psalm 145:21 similarly declares that "all flesh will bless his holy name forever and ever." These verses convey that a group referred to as "all flesh" will offer worship and reverence to God. I propose that animals are included in this group of "creaturely flesh," where their praise is distinct from that of inanimate beings and that all sentient human and non-human animals are capable of vocalising their praise of God.[92]

The book of Revelation extends this vision of universal worship to the eschatological future, situating creaturely praise within the heavenly realm: "Then I heard every creature in heaven and on earth and under the earth and in the sea, and all that is in them, singing, 'To the one seated on the throne and to the Lamb, be blessing and honour and glory and might, forever and ever!'" (Rev 5:13). This passage richly depicts the universal act of worship expressed by all living beings—those in heaven, on earth, beneath the earth, and in the sea—emphasising the collective praise directed towards God.

In the context of biblical praise, the elaborate songs of humpback whales can be seen as a form of animal praise arising from a state of flourishing in their environment. Research has shown that humpback whale songs are highly complex and may serve multiple functions, including communication and maintaining social contact during migration. These vocalisations, often occurring in feeding as well as breeding habitats, reflect the whales' thriving within their ecosystems.[93] Just as animals in biblical texts are called to offer praise to God, the vibrant and intricate vocalisations of whales can be seen as a natural expression of their existence and well-being. Since first being characterised as song nearly half a century ago, the cyclic and hierarchically structured vocalisation sequences of male humpback whales have become a significant subject in

[92] The concept of "all flesh" will be explored in greater detail in Chapter 4, where its theological and pneumatological implications are examined more comprehensively.

[93] Ryan et al., "Humpback Whale Song Occurrence Reflects Ecosystem Variability in Feeding and Migratory Habitat of the Northeast Pacific."

behavioural ecology research.[94] The complexity and variation of humpback whale songs have sparked extensive investigation into their functional roles and temporal evolution. Research indicates that humpback songs play a crucial role in both intersexual and intrasexual communication. It has also been hypothesised that these songs help maintain social contact during migration and facilitate echoic perception. Beyond these immediate functions, the evolution of humpback songs across vast spatial and temporal scales highlights their role in cultural transmission within and between whale populations. Favourable foraging habitats supported dense whale populations and enhanced density-dependent influences on song development and learning.[95]

While the songs of whales exemplify praise through flourishing, the biblical narrative also addresses the reality of animal lamentation and offers a holistic view of animal expressions in response to God. This duality of praise and lament will be explored further in the following section, where I examine how animals' interactions with God are depicted in scripture.

In the biblical writings, various passages (Is 43:20, Micah 1:8, Job 30:28–29, and Is 13:21–22) illustrate a vision of harmony between humans and animals in their expressions of grief and gratitude. For instance, in Isaiah 43:20, the wild animals' cries of lament in the wilderness—marked by environmental degradation and drought—are transformed into a harmonious celebration as God provides water for them:

> The wild animals honour me, the jackals and the owls, because I provide water in the wilderness and streams in the wasteland, to give drink to my people, my chosen…

This imagery suggests a cross-species connection where both humans and animals might share in a collective experience of rejoicing and gratitude.[96] Similarly, in Jonah 3:7–8, the king of Nineveh calls for both people and

[94] Ryan et al.

[95] Ryan et al.

[96] Moltmann refers to this interdependence as "symbiosis," a state in which all created beings exist in relation to one another, rather than in isolation, and are thus mutually dependent. Moltmann, *The Spirit of Life*, 219.

animals to participate in an act of repentance—fasting, donning sack-cloth, and earnestly appealing to the divine. This collective act highlights a shared sense of vulnerability and a potential for animals to be part of spiritual practices alongside humans. These examples reflect a broader theme of shared spiritual practices, where the experiences and expressions of animals are woven into the spiritual and communal life of humans, suggesting a vision of unity and shared spiritual experiences.

Hosea 4:3 presents a vivid image of ecological devastation, where "the land mourns and all who live in it languish; together with the wild animals and the birds of the air, even the fish of the sea are perishing." This text, which attributes the environmental crisis to a lack of faithfulness, kindness, and knowledge of God, illustrates the interdependence of all creation. It affirms that animals, along with humans, suffer the consequences of human wrongdoing, drawing attention to their vulnerability within the broader environmental context.

Something similar is reflected in the empirical research on elephants, which offers valuable insights into the emotional and cognitive capacities of animals, especially regarding their behaviours of mourning and lamentation. Observations by Pokharel, Sharma, and Sukumar document thanatological responses in Asian elephants, which exhibit behaviours indicating grief and mourning over deceased members of their social groups.[97] According to the authors, similar behaviours are found in species with strong social bonds and advanced cognitive capacities. The most common behaviour observed is the touch response, whereby elephants interact with the deceased by touching various body parts using their trunk and legs. This includes actions like trunk rest, tapping, shaking, and mounting-like postures, especially by young elephants. Tactile communication is crucial within elephant societies, as it helps them gather information about the well-being of others. Similar touch reactions are observed in non-thanatological contexts, such as attempts to invigorate newborn calves or consoling gestures during stressful events. Additionally, exploratory behaviours like investigating corpses and approaching ailing individuals are frequent. Changes in body postures, such as spreading ears and raising tails, suggest alertness and agitation among herd members near dead individuals. Epigenetic responses involving supporting or lifting dying or dead individuals are documented in various social mammals,

[97] Pokharel, Sharma, and Sukumar, "Viewing the Rare through Public Lenses."

including elephants, reflecting the kin-bonded social structure that motivates them to care for their ailing and deceased counterparts. Other observed behaviours include guarding the corpse, displaying aggression towards intruders, and reluctance to leave the deceased. These responses may serve to protect the corpse from predators or scavengers. The study also identifies different vocalisations and non-vocal signals produced by elephants in response to death, such as trumpets, roars, chirps, rumbles, trunk bounces, and ear-flapping. While the specific functions of these signals in the context of death remain to be fully explored, they underscore the complex emotional and social dynamics of elephants.

The scriptural witness to animals mourning and lamenting, when placed alongside empirical observations of species such as elephants who grieve their dead, unveils a profound truth: the emotional and cognitive lives of animals run deeper than often assumed. Such insights press against anthropocentric boundaries, suggesting that experiences of sorrow and lament are not the preserve of humanity alone. To see animals as capable of mourning is to glimpse them as participants in the spiritual and emotional fabric of creation. This broader vision reshapes theological and ethical imagination, calling us to re-evaluate their place within God's purposes and to approach all living beings with greater reverence. By recognising the depth of their feeling and the ways they too may respond to divine action, we are invited into a more expansive, respectful account of life—one that honours the interwoven griefs and hopes of the whole creation.

Prophetic speech serves as a conduit for communicating God's nature, while also drawing the world into God's divine life. It elicits a response that is itself a gift of the Spirit. This process of communication and response is grounded in the material and creaturely world, while also bearing a spiritual dimension. This section briefly explores how prophetic communication and response, guided by the Holy Spirit, are bestowed as a gift that encompasses all creation—including animals.

For Rachel Muers, the "cry of the earth" that is heard in the context of our environmental crisis can be seen as the Holy Spirit re-performing the gift of prophetic voice. Muers explains that non-human nature speaks with the power of the Spirit, sounding "prophetic laments and calls for repentance."[98] In her view, the Spirit summons the ordinary, creaturely

[98] Muers, "The Holy Spirit, the Voices of Nature and Environmental Prophecy," 333.

communicative capacity of the natural world into specific acts of communication that interrupt and transform history. Drawing on scriptural narratives of Spirit-given prophetic speech—from the prophets to Pentecost—Muers suggests that non-human nature, though it cannot literally speak, can nevertheless "make sense" in ways that are publicly available and comprehensible. This communicative capacity is not merely a projection of human meaning-making but arises from the inherent properties and relationships within the natural world itself. For Muers, creation's cry, filled with the Spirit, can sound prophetic laments, pronounce judgement, warn of consequences, and call for repentance and hope. She underscores that this speech not only communicates facts but produces transformative effects, enabling a response that is itself a gift of the Spirit. This understanding aligns with the broader Pentecostal and prophetic traditions where hearing and response are as integral to the Spirit's work as the speech itself, pointing to the Spirit's role in drawing creation into the life of God.

John Griffiths provides a biblical framework that supports this idea. According to Griffiths, Luke's redaction of Acts 2:19–20 suggests that the cosmic imagery in Joel 2:28–32 portrays a prophesying creation. Luke links these ecological wonders and signs to the ministry of Jesus and the apostles by modifying Acts 2:19, adding the term "signs" to Joel's language of wonders, creating the couplet of "signs and wonders" that parallels the "signs and wonders" associated with Jesus and the apostles. God is portrayed as the one who will reveal these signs, just as God worked signs and wonders through Jesus and the apostles during their earthly ministry. For Griffiths, Luke indicates that creation, like Jesus and the apostles, also performs signs and wonders through the Spirit of God as part of its prophetic calling and empowerment.[99] In another study, Griffiths probes further into the concept of creation prophesying, specifically exploring the utilisation of LXX Joel 3:1–5a by Peter in the book of Acts to clarify the outpouring of the Spirit during Pentecost in Acts 2.[100]

[99] John D. Griffiths, "Spirit-Baptised Creation: Locating Pentecost in the Meta-Narrative of Creation and Its Implications for a Pentecostal Ecology." *Australasian Pentecostal studies* 22, no. 1 (2021): 56.

[100] For Griffiths, the language used in Joel 3:1-5a implies the "all flesh" of all living creatures. Griffiths, John D. "Upon All Flesh: An Ecological Reading of Acts 2.17–21." *Journal for the Study of the New Testament* 47, no. 3 (March 1, 2025): 439. https://doi.org/10.1177/0142064X241301095.

He explains that the recurrence of the phrase "and they will prophesy" in Acts 2:18 alters the poetic structure of LXX Joel 3:1–5a from a chiastic arrangement to a direct parallelism that incorporates creation as a beneficiary of the Spirit. This alteration suggests that Lukan Peter perceived creational signs and wonders as forms of prophecy. The repetition of the phrase in Acts 2:18 does not augment the content of the passage but rather contributes to the overarching meaning of the outpouring of the Spirit, which leads to human prophetic manifestations and creational signs and wonders as a prophetic indication of the outpouring of the Spirit. As such, in Acts 2, Lukan Peter modifies the shape of LXX Joel 3:1–5a to encompass the notion that the heavens and the earth prophesy through non-verbal means.[101]

The perspectives of Muers and Griffiths draw attention to the prophetic weight carried by creation itself, where non-verbal signs and wonders emerge as testimony. The cries of suffering animals, in particular, echo as prophetic calls—voices that not only press upon the conscience of the church but also stir movements beyond its walls.[102] Such recognition cautions us against confining prophecy to human speech alone. Prophecy

[101] Griffiths, "Upon All Flesh," 446.

[102] While animals and the broader creation may participate in a form of prophetic expression through their actions, presence, and environmental interactions, the task of interpreting these prophetic voices rests with humans, endowed with the cognitive and spiritual faculties to discern meaning within complex ecological and relational contexts. The prophetic speech of animals, as described by Rachel Muers, emerges as "prophetic laments" that challenge human stewardship, calling for a response to creation's suffering. This lament, however, requires human listeners to engage deeply and empathetically with the signs around them, discerning spiritual and ethical implications within the natural world's signals. Muers, "The Holy Spirit, the Voices of Nature and Environmental Prophecy," 333.

To interpret the prophetic voices of animals, humans might need to cultivate a spiritual attentiveness and humility that acknowledges the limits of anthropocentric perspectives. By listening to creation with a posture of reverence and empathy, humans can begin to interpret the environmental "signs" as divine prompts for action. This approach aligns with John Griffiths' framework, where he describes how Lukan theology views non-verbal signs and wonders in nature as prophetic. This interpretive framework invites humanity to recognise the Spirit's activity within creation and to approach these signs with a theologically informed understanding that they are indeed part of God's revelatory work. In practical terms, humans can interpret animal prophecies by paying attention to ecological disruptions and the behaviour of animals as indicative of larger issues. For instance, the distress of marine life due to pollution, symbolised as the mournful "song" of whales, or the koalas' lament amid the ravages of bushfires, can be seen as calls for repentance and ecological care. Here, ecological science and environmental ethics become instrumental,

is not the sole domain of humanity; it resounds through the whole of creation, where even the voiceless bear witness to God's truth.[103]

Conclusion

In this chapter, I have sought to widen our vision of creation by tracing the Spirit's presence as it moves through the whole of life, particularly in the non-human world. The biblical language of *ruach* and *nephesh* opens a doorway here, reminding us that breath and life are not the exclusive possession of humanity but the shared gift of every living creature. To speak of the Spirit, then, is to speak of a creation alive with divine presence, where animals are not simply biological organisms but participants in a Spirit-drenched reality. They live, feel, and act out of an ontological depth that bears traces of sentience, even of spiritual responsiveness. Such a view stretches the boundaries of pneumatology, pressing us beyond narrow anthropocentrism towards a recognition of animals as integral to God's sustaining and relational work.

Alongside these theological reflections, I have engaged empirical studies that reveal the richness of animal life—their capacity for joy, grief, attachment, and communication. These insights from science do not stand apart from theology but, when brought into dialogue, sharpen our understanding of the Spirit's creative work. Animals, I suggest, may carry their own ways of expressing spiritual life, their own participation in creation's covenant with God. This recognition unsettles us: it asks whether the voices of creation have too often been muted by human-centred theologies and whether the Spirit might be drawing us towards a more expansive hearing.

Such an approach reshapes how we think about ethics and relationship. If the Spirit animates all creatures, then compassion and respect are not optional virtues but marks of faithful stewardship. Yet the picture is not without tension. For even as creation is suffused with Spirit, it is

serving as tools to decode the "language" of creation's suffering. Thus, human interpretation of the prophetic voices of animals requires a synthesis of scientific insight, theological reflection, and moral commitment. By drawing from both ecological data and a theology of creation, humanity can engage with these signs as meaningful and urgent messages from the Spirit, calling for transformative actions in the stewardship of the Earth.

[103] Fretheim would argue that God calls animals "in the service of the divine purposes; at the least we could speak of the nonhuman as both victim and vehicle."Fretheim, *God and World in the Old Testament*, 279.

also marked by predation, suffering, and loss—realities that refuse easy harmonisation. This complexity must be faced honestly if our theology of creation is to be credible. The next chapter takes up this task, turning to the question of predation and the difficult interchange between flourishing and death in a Spirit-filled world. Only by holding together both beauty and brokenness can we begin to discern the full breadth of creation's theological witness.

CHAPTER 3

Predation and Sin

When reflecting on sin, suffering, moral agency, and natural suffering, it is crucial to broaden theological discourse to encompass the experiences of non-human animals. Through a pneumatological lens, we gain a deeper insight into the Spirit's engagement with all of creation, especially the often-overlooked lives of non-human beings. As sentient creatures, animals navigate the complexities of suffering and may even exhibit rudimentary moral agency.[1] Including their experiences deepens our theological comprehension of sin and suffering as pervasive realities that extend beyond human agency, impacting the entirety of creation. By examining how the Spirit is present in, works through, and seeks to redeem the natural world, including its non-human members, this approach aims to engage with key themes that shape and enrich theological discourse.

The issue of natural suffering, particularly as manifest in predation, presents a significant theological challenge, requiring a confrontation with the complexities of creation, informed by both biblical narratives and scientific understandings of the natural world. While this chapter does

[1] The question of whether non-human animals possess moral agency is a complex and debated issue within theological and philosophical discourse. While this chapter engages a range of perspectives on animal moral agency, my own position is that non-human animals are not moral agents in the theological sense and therefore are incapable of sin. Their actions, while often complex and socially meaningful, do not involve the kind of moral deliberation and culpability attributed to human beings.

D. Rizzo, *Animal Pneumatology*, The Palgrave Macmillan Animal Ethics Series, https://doi.org/10.1007/978-3-032-18723-9_3

not aim to resolve the problem of predation, it acknowledges it as a vital issue requiring careful consideration within theological reflection.[2]

Natural suffering may be understood as any form of undeserved harm experienced by a self-conscious being that arises independently of human agency. Stephen Webb offers a helpful framework for discerning its character by setting it in contrast to human evil. Human evil, he argues, is marked by three elements: a perpetrator who acts beyond the bounds of relational or rational self-interest; an act that violates our shared sense of human dignity; and victims whose suffering demands justice.[3] Natural suffering lacks the first of these elements—there is no intentional agent behind the harm—yet it retains the latter two: injurious events and vulnerable victims.[4] Typically, discussions of natural suffering centre on human experiences of disease, natural disasters, or predatory attacks. But to draw a sharper line between natural suffering and human evil, we must widen the lens to include non-human animals. Their capacity for pain, their

[2] This chapter approaches the issue of suffering from an evolutionary perspective, aligning with many of the scholars engaged throughout this section. By adopting an evolutionary framework, I acknowledge the significant influence of Darwinian insights on contemporary theological reflection. The integration of scientific considerations into the discourse on natural suffering represents an important development that was unavailable to scholars prior to Darwin. By situating this discussion within the broader context of evolutionary biology, I aim to enrich the theological conversation by accounting for the natural processes underlying suffering (particularly in relation to animals) and thereby offer a contribution that is both nuanced and scientifically informed. Moltmann's pneumatological framework also incorporates insights from the natural sciences to highlight nature's orientation towards the future. In his ecological doctrine of creation, Moltmann describes evolution as a fundamental aspect of the "self-movement of the divine Spirit of creation." Moltmann, *God in Creation (The Gifford Lectures, 1984–1985)*, 19.

According to Moltmann, the Spirit serves as the guiding principle of evolution. He distinguishes between "creation" and "evolution," viewing the former as the miraculous emergence of existence itself, and the latter as the ongoing development of matter and systems of life. Moltmann, 196.

[3] Stephen H. Webb, *Dome of Eden: A New Solution to the Problem of Creation and Evolution* (Eugene, OR: Cascade Books, 2010), 92–93.

[4] Natural suffering encompasses events such as earthquakes and natural disasters, where no intentional action or deliberate agent can be identified. This lack of agency challenges attempts to attribute responsibility or moral intent to such occurrences. Consequently, natural suffering must be approached differently from human evil, which typically involves intentional actions that cause harm. The absence of agency in natural suffering invites a framework that acknowledges the occurrence of harm and the vulnerability of those affected, without implying moral culpability or intent behind these events.

awareness of suffering, and their potential both to endure and inflict harm make them indispensable to a fuller theological account.

One fundamental question is how we approach the Bible—whether from a literalist, creationist perspective or a position informed by evolutionary science. A science-informed reading of Genesis reveals a reality in which predation and death were present long before any notion of human sin. This raises considerable tension with the traditional depiction of an idyllic creation—a landscape where all creatures were vegetarian and free from suffering. If predation was integral to the evolutionary history of life on Earth, then it challenges the idealised vision of an Edenic creation and prompts a rethinking of how death and ecological dynamics have always been woven into the fabric of the natural order.

The presence of predation also calls into question the categorisation of such behaviour as "sinful."[5] In evolutionary terms, predation serves as a vehicle for ecological balance, facilitating diversity, population control, and the flourishing of ecosystems.[6] The natural interdependence and cycles of life and death challenge simplistic moral frameworks that might regard all suffering as inherently evil. The need to confront these realities becomes even more pressing when one considers the implications from a pneumatological perspective—how might the Spirit be understood to be at work in a world where death and predation are necessary for life?

In this chapter, I address the problem of natural suffering through a pneumatological framework, focusing on how the Spirit is present and active within the evolutionary processes and ecological dynamics that shape animal life. Within this view, the Spirit is understood not only as the Giver of life but also as the sustaining presence that upholds the complex relationships that constitute creation, including those between predator

[5] I adopt Ernst Conradie's redefinition of sin as a form of social diagnostics, viewing sin not as individual moral failings but as structural alienation from God, manifesting in ecological harm, inequality, and social injustice, and framing it as a broken relationship with the triune God. Ernst M. Conradie, *Redeeming Sin?: Social Diagnostics amid Ecological Destruction*, Religious Ethics and Environmental Challenges Series (Lanham: Lexington Books/Fortress Academic, 2017), xxv.

[6] Holmes Rolston contends that the emergence of life and the accompanying potential for new forms of existence inherently introduce new limitations. He argues that not even God can create a world where sentient beings can exist without the possibility of experiencing harm. This, he suggests, is both a logical and psychological impossibility, as the capacity for sentience necessarily involves vulnerability to pain and suffering. Holmes Rolston, "Redeeming a Cruciform Nature," *Zygon: Journal of Religion & Science* 53, no. 3 (September 2018): 742, https://doi.org/10.1111/zygo.12428.

and prey. I do not intend to diminish the brutal and often painful realities experienced by animals; these experiences must be acknowledged with full theological and moral seriousness. Yet, rather than interpreting predation solely as a destructive or regrettable necessity, this framework considers it as part of the Spirit's work in maintaining ecological balance. The Spirit's involvement encompasses both the flourishing of life and the intricate patterns of life and death that sustain ecosystems. In this light, the Spirit's redemptive mission extends to the entire created order, including the animal world, orienting creation towards its ultimate renewal. By attending to the Spirit's presence within predation and other natural processes, this chapter proposes a theological vision in which even the more difficult aspects of animal existence are situated within the Spirit's sustaining and transformative work.

This inquiry will unfold in several parts. First, the chapter will consider natural suffering (defined here as harm not caused by human agency) and the complexities it presents, particularly in relation to predation. This problem will be articulated, not to offer a resolution, but to highlight it as an essential aspect of theological discourse that requires careful reflection. This consideration serves to emphasise that suffering is not restricted to human experience but is an ecological reality with which the Spirit engages.

Next, I will consider moral and non-moral agency in animals and will draw on debates as to whether animals can be considered capable of moral behaviour. Considering the moral status of animals enables a deeper appreciation of the Spirit's interaction not only with morally accountable beings but also with those whose behaviours lie outside human moral categories. This discussion will then serve as a precursor to considering how these elements inform a pneumatological understanding of natural suffering and predation.

Finally, I will narrow its focus to a pneumatological interpretation of predation, considering in detail how the Holy Spirit may be understood to engage with this natural process. Particular attention will be given to the Spirit's relation to non-human animals within this context. I will propose a reconsideration of how predation might be integrated into a theological framework that views the Spirit's sustaining presence as active in the dynamic interplay of life and death in creation. Rather than attempting to resolve the problem of predation and suffering, I will highlight several key insights from a pneumatological perspective and will emphasise the Spirit's role in sustaining, guiding, and transforming creaturely life within a violent reality. These reflections will consider the implications for animals, whose experiences of predation and survival are deeply intertwined with the Spirit's creative and redemptive work in the natural world. The discussion will explore the Spirit as the sustainer of animal life, the Spirit's role in guiding creation towards fulfilment despite predation, the Spirit's creative dynamism within predation, and, finally, the Spirit's redemptive presence amidst the realities of predatory processes.[7]

Rather than assuming a definitive resolution, the aim here is to foster a nuanced appreciation for the Spirit's role in sustaining and guiding a creation marked by both beauty and brutality, life and death, suffering and hope. The challenge lies in holding these tensions in creative dialogue, recognising that even within the harsh realities of predation, the Spirit of God may be discerned as intimately present—sustaining animal life, bearing witness to the vulnerability and suffering that characterise much of the natural order, and pointing towards an ultimate redemption of all creatures.

I make no claim to address the full breadth of theodicy, a question that has occupied theology for centuries. My aim is more modest: to reflect on natural suffering (especially predation) by considering how pneumatology might open new ways of thinking about this difficult reality. The intent here is not to justify or explain away the suffering that exists within creation but to explore how the Spirit's active presence might be perceived amid the complexities of life and death that permeate the animal kingdom. In this way, the chapter invites a deeper awareness of the Spirit's redemptive and sustaining work within the lives of animals that eagerly anticipates its final healing and wholeness.

[7] The question of whether the Spirit is aligned with the predator or the prey introduces a complex theological tension. While creation is indeed affirmed as "good" (Gen 1:31), it nonetheless "groans" in anticipation of redemption from suffering, decay, and death (cf. Rom 8:19–22). This eschatological vision does not negate the goodness of the created order but rather points to a future where the realities of predation and suffering are ultimately transformed within the fullness of God's peace. In this vision, the Spirit's presence may be understood as both sustaining the current dynamics of life, including predation, and simultaneously orienting creation toward its final reconciliation and the promise of a new creation, where suffering and conflict are no more. This will be addressed later in chapter 5.

Considering Natural Suffering

In addressing the problem of natural suffering, particularly with regard to the animal kingdom, Ernst Conradie considers the philosophical tension between evolutionary mechanisms and theological interpretations of creation. Central to this tension is the dynamic between the flourishing and the elimination of diversity. Evolution, through processes such as mutation, hybridisation, and natural selection, engenders survival by amplifying variation, yet it concurrently narrows possibilities, discarding those deemed weak or fruitless.[8] This process not only unfolds at the level of individual organisms but also extends to entire species, ecosystems, and even cultural and intellectual paradigms. Conradie notes that the majority of living creatures—both human and non-human—fail to reach full maturity. This raises deeply unsettling theological questions about the nature of divine creation.

The stark realities of natural selection, exemplified in the premature death of many creatures to ensure the survival of a select few, challenge the coherence of the idea of a benevolent Creator. Conradie interrogates whether such a system reveals a God aligned with the strong rather than the weak. This question forces a confrontation with the larger theological problem of theodicy: how can a world teeming with suffering and death be reconciled with the biblical pronouncement of creation's inherent goodness?[9] This problem raises the question of whether natural suffering—manifest in predation, suffering, and mortality—is a byproduct of a fallen creation or an inherent characteristic of an unfallen world. Conradie's reflections grapple with how a Creator, understood as both loving and just, can be reconciled with evolutionary processes that seem to prioritise the strong at the expense of the weak. This tension sets the stage for a deeper theological exploration of natural suffering, particularly within the framework of whether creation is to be understood as fallen or unfallen.[10]

This section considers the idea that predation and natural processes, including suffering and death, are not the result of a cosmic fall but are part of the original and intended natural order of creation. It challenges the traditional view that the world has been fundamentally corrupted by

[8] Conradie, *Redeeming Sin?*, 200.

[9] Conradie, 201.

[10] Conradie, 202.

human sin, suggesting instead that creation, including the non-human animal, remains unfallen in its natural state.

Bethany Sollereder challenges the conventional narrative of the fall as an explanation for predation and other forms of natural suffering.[11] She outright rejects the notion that the natural world, apart from human sin, is fallen.[12] Sollereder maintains that the natural world is not corrupted by human sin or any primordial fall, but rather aligns with God's purposes, even if it is not the best of all possible worlds. Evolution, in her view, though filled with disvalue and suffering, is an authentic expression of divine love.[13] I find this line of reasoning compelling. Her argument

[11] Sollereder, Bethany N. *God, Evolution and Animal Suffering: Theodicy Without a Fall* (London/New York: Routledge, 2019). Sollereder's work is engaged here primarily due to her focused attention on animal suffering within the broader discourse of natural suffering. Unlike many theologians who centre their discussions on human sin and redemption, Sollereder uniquely addresses the theological implications of non-human suffering, particularly in relation to predation. Her rejection of the traditional Fall narrative as the cause of predation aligns with my interest in exploring how animal suffering can be understood within the framework of divine love and creation's inherent goodness. Sollereder's approach offers a compelling lens through which to examine the place of animals in creation, emphasising their participation in evolutionary processes without resorting to notions of a corrupted or fallen world.

[12] Other scholars who align with this position include Christopher Southgate, Elizabeth Johnson, Michael J Murray, Jon Garvey and Ronald Osborn who similarly argue that natural processes involving predation and suffering are part of God's intended order rather than a consequence of a corrupted creation. See Christopher Southgate, *The Groaning of Creation: God, Evolution and the Problem of Evil* (Louisville, Kentucky: Westminster John Knox Press, 2008); Ronald E. Osborn, *Death before the Fall: Biblical Literalism and the Problem of Animal Suffering* (Downers Grove, IL: IVP Academic, 2014); Garvey, *God's Good Earth: The Case for An Unfallen Creation.*; Elizabeth A Johnson, *Ask the Beasts: Darwin and the God of Love* (London, UK: Bloomsbury Continuum, 2014); Michael J Murray, *Nature Red in Tooth and Claw: Theism and the Problem of Animal Suffering* (Oxford, UK: Oxford University Press, 2008). For Southgate, there is no historical fall from an initial state of perfection. Instead, he views the natural world, including all its inherent disvalue, as part of God's intended creation, making evolutionary suffering a necessary condition for the emergence of complex beings. While Southgate rejects the concept of a historical Fall and sees suffering as an integral part of the evolutionary process ordained by God, Conradie stresses the need to distinguish between natural suffering and social evil. Conradie argues that the narrative of salvation should not only address natural suffering but also emphasise human responsibility for social evil and its harmful effects on the environment and other living beings. See Ernst M. Conradie, "On Social Evil and Natural Evil: In Conversation with Christopher Southgate," *Zygon: Journal of Religion & Science* 53, no. 3 (September 2018): 752–65, https://doi.org/10.1111/zygo.12427.

[13] Sollereder, *God, Evolution and Animal Suffering,* 184.

that predation is not a deviation from God's creative intent but an integral feature of creation provides a crucial theological reframing. By recognising evolutionary processes (including predation) as instruments through which God brings about complexity and flourishing, Sollereder rightly unsettles traditional theodicies that tether natural suffering to sin or rebellion.[14] This position resonates with my own conviction that the dynamics of life and death within creation, however difficult, must be taken seriously as part of the Spirit's sustaining and life-giving work.

Sollereder critiques attempts to attribute the disordered aspects of creation to either human sin or an angelic rebellion. Historically, theologians like John Calvin have asserted that all forms of natural disorder (whether disease, frost, or predation) are consequences of sin. Sollereder, however, insists that creation, as it stands, fulfils God's purposes, though this does not mean it is the best of all possible worlds or that individual suffering can always be justified within this framework. Instead, she describes evolution as a process full of "disvalue" that is nevertheless a genuine act of divine love. Sollereder aligns with thinkers such as Paul Fiddes and Elizabeth Johnson in depicting divine love as intrinsically self-giving—a love that involves God's willing entry into a vulnerable relationship with creation.[15] Fiddes' claim that God's suffering is wholly self-chosen underscores a divine freedom that is neither overwhelmed nor constrained by creation but actively and lovingly engaged with it. This understanding reinforces Sollereder's rejection of the notion of a corrupted natural world.

[14] Arthur Peacocke's assertion that pain and suffering are inherent and necessary elements of biological evolution aligns with the views of Sollereder, who similarly recognises these phenomena as integral to the survival and development of life forms. Both scholars reject the notion that such suffering arises from human moral failings or "sinfulness," instead framing it as a fundamental part of the natural world. This alignment highlights a shared understanding of the evolutionary process, where human beings, despite their heightened awareness, are not the originators of suffering but rather experience it as part of a broader ecological reality. Arthur Peacocke, "The Cost of New Life," in *The Work of Love: Creation as Kenosis*, ed. John C. Polkinghorne (Grand Rapids, MI: Eerdmans, 2001), 31–32. However, in this view, the idea that God might have designed predation challenges deeply ingrained assumptions about the incompatibility between a loving Creator and a world characterised by suffering and death.

[15] See Paul S Fiddes, *The Creative Suffering of God* (Oxford: Clarendon Press, 1991); Johnson, *Ask the Beasts*, 191.

While Sollereder reframes natural suffering within the logic of divine love, I find it equally compelling to draw on John Polkinghorne's reflections on the continuity between the old and new creation, which deepen this theological conversation.[16] Polkinghorne insists that the new creation is best understood not as an entirely novel act of divine fiat but as a redeemed transformation of the existing order. Drawing on the model of Christ's resurrection—where the risen body represents both continuity and transformation—he proposes that the new creation emerges *ex vetere* (out of the old), rather than *ex nihilo* (out of nothing). This presses us to consider why God allows a world marked by suffering to unfold if a pain-free existence ultimately awaits. For Polkinghorne, it is not divine negligence but rather a twofold creative strategy: first, God brings the world into being and allows it to develop freely at a metaphysical distance; then, through Christ's redemptive work, the world is drawn into an intimate, transformed communion with God.[17]

Jeff Astley extends this line of thought by challenging the assumption that divine care is only manifest when suffering is prevented or alleviated. He describes the Spirit as a "metaphysical whisper" underpinning all natural processes, a quiet but constant presence that sustains creation even in its most brutal moments.[18] This pneumatological vision recognises the Spirit's active presence within the cycles of predation, illness, and death—realities that shape the lives of animals as much as humans. I find this image of the Spirit as a sustaining thread of divine love particularly compelling. It neither denies the harshness of animal existence nor romanticises it; rather, it locates God's presence within the fabric of creaturely life, weaving love through both its beauty and its pain.

For animals, whose lives are marked by their own experiences of suffering and the struggle for survival, the Spirit's presence affirms their

[16] John Polkinghorne, *The God of Hope and the End of the World* (New Haven, CT: Yale University Press, 2002), 116, http://ebookcentral.proquest.com/lib/harvest/detail.action?docID=3420227.

[17] Celia Deane-Drummond would critique Polkinghorne (as she does Arthur Peacocke) by arguing that framing suffering as necessary risks inadvertently justifying its existence. See Celia Deane-Drummond, *Christ and Evolution: Wonder and Wisdom* (Minneapolis: Fortress Press, 2009), 171–72.

[18] Jeff Astley, "Evolution and Evil: The Difference Darwinism Makes in Theology and Spirituality," in *Reading Genesis After Darwin*, ed. Stephen C Barton and David Wilkinson (Oxford, UK: Oxford University Press, 2009), 171.

importance within the broader creation. It is a vision of divine engagement that includes their experiences and suggests that they are not abandoned in the face of natural suffering but are upheld by the same divine love that sustains all beings. By reframing the notion of divine absence in the midst of animal suffering (rather than seeing the lack of miraculous intervention as a sign of indifference) amplifies the Spirit's persistent and loving presence in and through all of creation.

An important consequence of this vision is that it sustains the importance of animals, as it provides the foundation for the future transformed state of creation. Thus, dismissing environmental care or neglecting the present world becomes untenable from a theological standpoint. Polkinghorne contends that his approach offers a more robust and realistic basis for affirming the worth of the present creation compared to speculative utopian visions that undervalue the significance of the existing world.

Christopher Southgate proposes that the evolutionary struggle of creation can be understood as the "travail" to which God has subjected creation, with the hope that the values of complex life, and ultimately freely choosing beings such as ourselves, might emerge.[19] Southgate

[19] Southgate, *The Groaning of Creation: God, Evolution and the Problem of Evil*, 95. In addition to this idea, Christopher Southgate engages with the theological challenge of why a perfectly good and all-powerful God would allow for a world that includes struggle, competition, and violence, particularly within an evolutionary framework. He draws on proposals by Neil Messer, who references Karl Barth's concept of *Das Nichtige* (nothingness), and Celia Deane-Drummond, who invokes Sergius Bulgakov's "Shadow Sophia.“ Both propose that there is a mysterious limitation on God's creative power, leading to the emergence of a world where creaturely flourishing is intertwined with struggle and suffering. This problem can be viewed from two perspectives. If the constraint on God's creative activity is seen as a spiritual force, this raises concerns about dualism, suggesting that God might be engaged in a cosmic struggle with contrary spiritual powers that affect the character of creation. This also risks undermining the classical Christian understanding of God as sovereign and the source of absolute goodness. On the other hand, if the constraint is understood as a logical necessity inherent within the nature of creation, theologians are faced with the challenge of explaining what this logic entails and why it limits God's ability to create a world where flourishing occurs without competition or violence. Southgate highlights the importance of grappling with this tension in order to preserve a coherent view of God's sovereignty and goodness while taking seriously the realities of evolutionary processes and the existence of natural suffering in the created order. This ongoing theological conversation seeks to reconcile traditional doctrines of divine power with the complexities of an evolving creation, where natural processes often involve struggle, competition, and suffering. These complexities prompt deeper reflection on how divine sovereignty and goodness can coexist with the reality of natural suffering.

views death as a thermodynamic necessity, which need not be considered problematic if it follows a life that has reached fulfilment and completion.[20] Southgate makes reference to the Apostle Paul's language and suggests that this evolutionary struggle resembles the process of giving birth—a dynamic movement of growth and transformation.[21] Within this travail, both human and non-human creation participate together, with humanity contributing through the laborious process of bringing forth new possibilities of life.

Taking a more pneuma-centric approach, Denis Edwards offers a compelling vision of God as continually present through the Word and the Spirit—breathing life into creation's unfolding processes moment by moment. God is not distant or detached but dynamically engaged, responsive, and intent on bringing forth a world of creatures capable of receiving divine self-giving love.[22] What stands out to me in Edwards's account is his portrayal of God's *creative patience*: a God who waits for creation to emerge in its own time, allowing evolutionary processes to unfold with all their complexity and unpredictability. This vision does not attribute deliberation or sentience to natural processes themselves, yet it depicts God as joyfully responsive to creation's unfolding, delighting in its independence and integrity.

Christopher Southgate builds on Edwards's insights by emphasising the Trinitarian generosity that makes space for creation's freedom. He reinforces the idea that creation is an act of divine love—an act in which the Persons of the Trinity willingly accept the limitations inherent in

Christopher Southgate, "Did the Natural World Go Wrong?," in *God and the Natural World: Theological Explorations in Appreciation of Denis Edwards*, ed. Ted Peters and Marie Turner (Adelaide: ATF Press, 2020), 63.

[20] Denis Edwards, *How God Acts: Creation, Redemption and Special Divine Action* (MPLS: Fortress Press, 2010), 13. Further to this, Southgate argues that creation, as it has evolved, is imbued with actual and potential beauty, creativity, and the flourishing of life. However, these values are inextricably linked with competition, suffering, and extinction—inevitable aspects of a Darwinian world. While the human world has experienced profound alienation, estrangement, and brokenness, the non-human world, despite its suffering, does not inherently reflect this same alienation or brokenness. It is only where human actions have caused irreversible damage that such brokenness becomes apparent in the natural world. Southgate, "Did the Natural World Go Wrong?," 67.

[21] Southgate, *The Groaning of Creation: God, Evolution and the Problem of Evil*, 95.

[22] Edwards, *How God Acts*, 66.

allowing creation to develop freely.[23] I find this particularly important because it reframes the theological narrative: God is not a frustrated architect unable to produce a flawless world, nor the cruel deity caricatured by some atheist critiques. Instead, God chooses to allow creation to flourish on its own terms, even when that involves struggle, predation, and pain.

Both Edwards and Southgate invite us to imagine a God whose engagement with creation is marked not by coercive control but by a self-giving love that sustains and accompanies. From a pneumatological perspective, this paints a picture of the Spirit as intimately woven into the very fabric of evolutionary becoming—an active, life-giving presence that honours creation's integrity (in its complexity and unpredictability) while holding it within the divine embrace.

I find in this account a compelling vision of the Spirit as the breath that animates and holds together a complex creation—a presence that does not stand apart from the messiness of evolutionary history but moves within it. The Spirit is not a distant force, intervening sporadically to correct nature's course, but the constant, life-giving presence that permeates even the difficult terrain of suffering, death, and predation. Through the Spirit, God accompanies creation on its long and often tumultuous journey, granting genuine autonomy to the natural order and making space for the emergence of complex, interdependent life.

This way of framing the Spirit pushes against the instinct to imagine divine action as a kind of supernatural override and affirms the independence and integrity of non-human creatures, acknowledging their participation in a creation that is both beautiful and fraught. Within the reality of natural suffering, the Spirit's role is not to prevent pain or to smooth away the rough edges of evolutionary processes, but to sustain and guide creation through them. What Edwards articulates here is a pneumatological response that neither romanticises nature nor turns away from its brutalities. The Spirit is present precisely in those ambiguous spaces—in the unfolding, in the groaning, in the persistent movement of life towards its promised renewal.[24]

[23] Southgate, "Did the Natural World Go Wrong?," 68.

[24] Furthermore, Edwards' portrayal of God delighting in the "emergence of a world of creatures with their own independence and integrity" offers a theological foundation for understanding animal life as possessing intrinsic worth. It suggests that animals, with their unique experiences of life—including suffering—are active participants in the divine creative process. The Spirit's involvement affirms the importance of non-human creatures

Taken together, these theologians offer complementary perspectives that align with and expand on themes similar to those explored by Sollereder: creation, including animal life with its inherent suffering and predation, is intrinsically valuable and operates within God's intended purposes. Sollereder's distinctive contribution lies in her unwavering claim that animal suffering, far from signalling a fallen or corrupted creation, is an essential element of God's evolutionary design.[25] Predation and death, rather than being consequences of sin, are part of a natural order that allows for complexity and flourishing in the created world. With this non-fallen framework established, I now turn to consider the concept of cosmic sin and how it intersects with these realities.

As I proceed, I will summarise key arguments that address the problem of natural suffering, particularly in relation to the idea of cosmic sin. By doing so, I articulate ways in which sin, predation, and suffering are interpreted within theological frameworks that attempt to uphold both the integrity of creation and the role of divine providence. The following section considers the theological view that animal predation and suffering in the natural world are consequences of a cosmic Fall. According to this perspective, creation has been negatively impacted by human sin, resulting in a world that is no longer in the state of harmony as originally intended by God.

Michael Lloyd presents a strong argument that predation is not part of God's original design for creation. Instead, he views it as a distortion introduced by a cosmic Fall, and suggests that the violence and suffering evident in the natural world are the result of this catastrophic event, rather than inherent to God's intended order. Lloyd critiques attempts to justify

within the broader narrative of creation, even amidst the realities of predation and death. This positions natural suffering not simply as a problem to be solved but as a part of the Spirit's ongoing, patient work in bringing about the fullness of creation, pointing ultimately towards a future hope of renewal where all creatures participate in the redemptive purposes of God. Edwards, *How God Acts*, 66.

[25] Sollereder, *God, Evolution and Animal Suffering: Theodicy Without a Fall*, 93. Sollereder's argument builds upon the foundational work of theologians such as Christopher Southgate, Arthur Peacocke, and Holmes Rolston, who have all addressed the role of suffering within an evolutionary framework. Her distinctive contribution lies in her commitment to viewing animal suffering not as evidence of a fallen creation but as an integral part of God's creative intentions. By engaging with these predecessors, Sollereder expands the conversation on the importance of creaturely life, including the complexities of animal existence within God's evolutionary design.

predation within God's creational purposes, maintaining that such arguments undermine the goodness and self-giving nature of God, as revealed in Christ.[26] Lloyd's analysis sees predation as incompatible with divine intent. He frames the problem of natural suffering as particularly challenging for Christian theology, as it calls into question the nature of God in light of the violence present in the natural world. This tension, Lloyd argues, can only be resolved by understanding predation as a result of a fallen creation. The comparison between Christ's self-giving sacrifice and the ruthless nature of animal predation brings into sharp relief the theological challenge of viewing predation as part of God's original plan.

In presenting the case for a fallen creation, Lloyd critiques views that minimise or deny the problem of predation by asserting that animals do not experience pain in the same way humans do. He writes: "Animals are sentient creatures capable of suffering and as such are a moral end in themselves."[27] He rejects dismissals of this view as oversimplifications that fail to grapple with the moral weight of observable suffering in the natural world. Lloyd further challenges instrumental views, which claim that predation serves a greater good, and argues that such approaches diminish God's goodness by making suffering an integral part of creation.[28] Instead, he asserts that suffering and predation

[26] See Michael Lloyd, "Are Animals Fallen?," in *Animals on the Agenda*, ed. Andrew Linzey and Dorothy Yamamoto (London, UK: SCM Press, 1998), 147–60. Lloyd frames the depiction of Christ's sacrifice on the cross as an ontological foundation for Christians' moral protest against the cruelty evident in nature, highlighting the inherent tension between the self-giving love of Christ and the "ruthless self-preservation" of animal predation.

[27] Lloyd, 151.

[28] Consequently, for Lloyd, animal behaviours characterised by violence and predation cannot be understood as reflective of God's original purpose. Lloyd identifies three key supports for this view. First, the Levitical laws concerning cleanness and uncleanness, which declared predatory animals unclean, suggest that predation falls outside the divine order of holiness. Second, the miracles of Jesus—calming storms, healing the sick, and raising the dead—demonstrate the re-establishment of divine control over a fallen creation. Third, prophetic visions of creation at peace point towards an eschatological hope for a restored creation, wherein predation and violence are ultimately undone. These three elements indicate that the destructive interactions observed in the natural world are contrary to God's creational intent and must be viewed as historically distorted.

are deviations from God's original intention, which will ultimately be redeemed and restored through Christ.[29]

Lloyd also critically engages with speculative theological proposals, such as instrumental views of suffering which complicate theological conceptions of God's character.[30] He believes that these blur the distinction between God allowing suffering as a consequence of freedom and God directly willing or creating suffering for a specific purpose.[31] He compares this to a parent who might allow a child to face difficult experiences for growth but would never deliberately create such experiences. This distinction is vital for preserving the image of God as genuinely benevolent. Furthermore, instrumental accounts that depict God as the author of suffering diminish the authenticity of divine praise and gratitude. If God is the creator of the very darkness from which humanity is redeemed, the sincerity of praising God as a redeemer is undermined. The Incarnation, Lloyd notes, is aimed at healing and redeeming a broken creation rather than rectifying a state of suffering deliberately intended by God. More compelling for Lloyd is the view that the fall of angels may provide a framework for understanding the disordered state of creation.[32] In this view, predation is seen as a symptom of the angelic fall, a consequence of spiritual rebellion that has distorted the natural world.

By firmly positioning predation as a manifestation of the fallen order, Lloyd insists that predation is incompatible with the nature of a loving God. His theological framework calls for understanding predation and suffering as deviations from God's original purpose for creation. Through

[29] Moreover, Lloyd argues that even in the absence of suffering, the inherent predation in the natural world appears incompatible with the image of God's glory as reflected in Christ's self-giving love. Attempts to justify predation as part of God's creative purposes, often adopt values that are fundamentally anti-Christian.

[30] To be clear, instrumental views of suffering propose that suffering serves a larger purpose or contributes to a greater good within the broader framework of creation. In this perspective, suffering is not seen as inherently negative but as a necessary element that brings about important outcomes, such as growth, moral development, or the flourishing of life.

[31] The instrumental approach also risks creating a theological dichotomy between creation and redemption. Lloyd emphasises that Christian soteriology affirms the continuity between creation and redemption—salvation is about restoring creation to its intended purpose rather than discarding or opposing it. By suggesting that suffering is essential to creation, the coherence between these two aspects of Christian belief is weakened, ultimately undermining the redemptive purpose of Christ.

[32] Lloyd, "Are Animals Fallen?," 159.

the redemptive work of Christ, this fallen state will ultimately be healed and transformed, offering hope for the restoration of creation to its intended harmony. While Lloyd's concerns about preserving the goodness of God are both important and pastorally sensitive, I do not share his conclusion that predation must be understood as a deviation from God's creative will. Such a view risks placing natural processes (integral to the emergence and flourishing of life) outside the scope of God's providential activity. It also reinforces a stark dualism between creation and fall that can flatten the complexity of the natural world. As Sollereder and others have argued, evolutionary processes, including predation, need not be read as evidence of corruption but can be understood as part of God's creative economy—difficult, yes, and at times brutal, but nonetheless caught up within divine purposes. In my view, Lloyd's framework, while theologically motivated, tends to cast creation's natural order as something to be undone rather than redeemed through transformation. This differs from a pneumatological vision in which the Spirit is actively present within and sustaining these processes, even as they are caught up in suffering and awaiting ultimate renewal.

Although these accounts diverge sharply in their theological interpretations, I find it fruitful to hold them in constructive tension rather than set them in opposition. Sollereder offers a compelling vision of predation and natural suffering as integral to God's creative design—a world shaped by divine love that nonetheless unfolds through complex and often difficult processes. In contrast, Lloyd underscores the cosmic Fall as a way of naming suffering and death as distortions within creation's harmony, directing theological attention towards future redemption and restoration. Rather than treating these perspectives as mutually exclusive, I see them as illuminating different facets of the theological task. Sollereder's approach anchors the reality of predation within God's creative purposes, while Lloyd's framework gives weight to the eschatological hope that suffering will one day be overcome. My own position leans towards Sollereder's account, yet I acknowledge the unresolved tensions that continue to animate theological reflection on natural suffering and predation.

The Moral Status of Animals

The question of whether predation and suffering are aspects of a fallen creation inevitably raises further theological questions about the moral status of animals—specifically, whether they can be considered moral agents, possess moral culpability, or bear the consequences of moral fallenness within such a framework. If, as Lloyd suggests, predation is not part of God's original design but rather a consequence of a cosmic fall, this invites inquiry into the extent to which animals might participate in moral dynamics within a creation that is, in some sense, disordered. This section thus builds upon the preceding discussion by examining whether certain animal behaviours—especially those involving violence or predation—could be interpreted as morally significant within a fallen order. Engaging perspectives that both challenge and support the notion of animal moral agency, this section will explore how traditional and contemporary theological accounts address the complex relationship between animal behaviour, moral responsibility, and the broader implications of living within a world affected by natural suffering.

Clough offers a distinctive approach to the issue of non-human moral agency. He challenges traditional theories by suggesting that certain animal behaviours—such as predation within established relationships—could be viewed as morally significant acts, akin to rebellion or sinfulness. He explains how, historically, non-human animals were sometimes treated as ethically accountable and faced consequences similar to those imposed on humans for immoral actions.[33] For instance, violent behaviour towards humans or engaging in prohibited relations could result in punishment by stoning (cf. Lev 15–16), and animals who did not fast during periods of repentance, as in Jonah 3, were held responsible for their disobedience. Rather than assuming that animals fully understand God's moral expectations, Clough argues that the question of moral responsibility is more nuanced. His primary scientific case study centres on a group of cannibalistic chimpanzees. In examining whether their behaviour—slaughtering and consuming the young of their own species—should be considered morally wrong, sinful, or neutral, Clough suggests that the ability of chimpanzees to make deliberate choices about their actions might be akin to that of a human child, rather than an adult.[34] However, just as we

[33] Clough, *On Animals, Systematic Theology*, 109.

[34] Clough, 112–13.

do not believe children instantly transition from automatons to morally responsible agents at a specific age, this gradation of moral judgement should not exclude chimpanzees from being viewed as morally culpable for sinful actions.[35]

Celia Deane-Drummond acknowledges a significant shift in her views regarding the moral capacity of animals. In her earlier work, she argued that morality is an exclusively human trait and contended that non-human animals should not be considered morally culpable in a strong sense.[36]

[35] Sollereder insightfully critiques Clough's position, arguing that it may be somewhat misleading to use the most mature members of the most intelligent non-human species as a case study to suggest that sin, manifesting as violence, extends beyond humanity. Clough's primary claim—that violence in general arises from non-human actions in opposition to God's will—faces challenges when applied across the entirety of non-human creation. While proto-moral behaviours may be observed in the most advanced sentient animals, attributing moral rebellion to creatures present during the early Cambrian period, when predation first appeared, is far more problematic. Furthermore, while Clough references biblical traditions that hold non-human animals morally accountable (e.g., the stoning of an ox that gores a human, cf. Exodus 21:28), these laws apply only to higher-order animals, with no equivalent for lower life forms, such as locusts. Sollereder also observes that Clough does not engage with passages where God actively provides for a violent creation, as seen in Psalm 104:21–27 and Job 39:30. Moreover, Clough's sources, which include the biblical narrative of Eden and medieval legal cases, predate any understanding of evolutionary biology, a significant omission in addressing the complexities of moral agency in non-human animals. Sollereder further critiques Clough's inability to reconcile the evolutionary processes that inherently link values and disvalues—an issue Clough himself acknowledges. His argument that all creatures are mired in sin is rooted in Colossians 1:20, which speaks of Christ reconciling "all things." If all creation requires reconciliation, Sollereder asks whether this extends to heavenly beings as well, questioning whether sin is present in heaven. If not, she suggests, there must be a way for non-human creatures to be reconciled to God through Christ without necessarily being implicated in sin. I will explore this connection further in the next chapter. See Sollereder, *God, Evolution and Animal Suffering*, 68–69. Veli-Matti Kärkkäinen also critiques Clough's argument regarding the moral responsibility of animals as presented in *On Animals*. Kärkkäinen challenges Clough's claim that animals can be morally responsible for their actions, arguing that such responsibility would require a level of self-reflection that animals are scientifically and theologically unlikely to possess. He views Clough's assertion as implausible, stating that moral culpability cannot reasonably be extended to non-human creatures, given their limited capacity for moral discernment. Veli-Matti Kärkkäinen, *Creation and Humanity: A Constructive Christian Theology for the Pluralistic World, Volume 3*, Illustrated edition (Grand Rapids, MI: Eerdmans, 2015), 176.

[36] It is important to distinguish between moral culpability and the theological concept of sin. As Conradie notes, equating sin with moral culpability risks reducing sin to purely ethical dimensions, thereby overlooking its relational aspects. In biblical and theological contexts, sin is understood primarily in terms of a disrupted relationship with God,

Recent research conducted by primatologists and animal behaviourists like Bekoff and de Waal has significantly shaped Deane-Drummond's perspective on animal morality. These studies have led Deane-Drummond to propose a new framework that recognises a form of animal morality that respects their behaviours and social worlds while helping to identify similarities and differences with human moral behaviour.[37] She notes that such a position may be challenging for some philosophers, who might view it as a categorical error, but only if one assumes that human morality is the only possible model for understanding moral or immoral relations. Here, Deane-Drummond emphasises the importance of recognising the differences between humans and non-human animals, asserting that animals cannot simply be absorbed into the human moral community.[38]

In her argument, Deane-Drummond insists that acknowledging non-human animal morality does not entail projecting human virtues or vices onto animals. Drawing on Mary Midgley's critique of crude anthropomorphism, Deane-Drummond distinguishes between simplistic projections of human attributes and the more nuanced approach advocated by ethologists, who seek to understand animal behaviour within the context of animals' own lives and social structures.[39] Addressing the theological dimension of this issue, Deane-Drummond explores the challenges theologians face when considering the notion of moral behaviour in non-human animals. She acknowledges that respected theologians, including Denis Edwards, remain uneasy with this idea—a sentiment she shared until engaging with recent ethological research. She suggests that while anthropomorphism should be avoided, excluding the possibility of animal morality outright is unwarranted.

By drawing parallels between the development of animal rights and the moral consideration of animals, Deane-Drummond stresses the need for adapting moral frameworks to accommodate non-human creatures. These considerations for animals should not be simple extensions of human

rather than solely as ethical failure. See Conradie, *Redeeming Sin?*, 130. This distinction is essential to avoid a moralistic or reductionist interpretation that confines sin to moral behaviours, overlooking its deeper, relational implications. Thus, while Deane-Drummond's exploration of animal moral capacity opens intriguing discussions about ethical responsibility, this should not be conflated with sin in the theological sense.

[37] Deane-Drummond, *Christ and Evolution*, 162.

[38] Deane-Drummond, 162.

[39] Deane-Drummond, 163.

rights. Instead, they should be adapted to the unique needs of non-human animals. Additionally, she proposes that there is an evolutionary continuity between animal and human moral behaviour.[40] Though humans have unique capacities, many of their behavioural tendencies are linked to their evolutionary forebears. She does not equate non-human moral capacity with human morality but acknowledges that moral and immoral tendencies are present in both human and non-human animals. Consequently, she speculates that non-human immoral behaviour, insofar as it involves an element of choice, might also stand in need of redemption. Deane-Drummond suggests that the roots of morality and virtue are not unique to human history but emerge within a broader evolutionary process. Observations of empathy and altruism in primates may reflect rudimentary forms of morality, acting as precursors to more complex ethical systems in humans. She advocates for a nuanced perspective that acknowledges both the admirable and limited aspects of animal behaviour, proposing that these behaviours reveal moral potential alongside inherent limitations within the non-human animal world.

While Deane-Drummond's engagement with ethology offers a valuable corrective to overly anthropocentric moral frameworks, I remain unconvinced that animals should be construed as moral agents in a way that implies culpability or sin. Behaviours that resemble human moral tendencies—such as empathy, cooperation, or aggression—can and should be interpreted within the contours of animal social worlds without imposing moral responsibility that presupposes a level of agency they do not possess. From my perspective, animals participate in creation as Spirit-animated beings, capable of remarkable relational and behavioural complexity, but not as moral agents capable of sin. Their lives are profoundly affected by sin's consequences—ecological disruption, suffering, and death—but they are not themselves the bearers of sin in a theological sense. Recognising this distinction is essential for constructing a pneumatology that honours animal life without overextending categories of moral agency that are properly theological-anthropological.

Ernst Conradie, in "Do Only Humans Sin?", engages critically with Frans de Waal's research on moral behaviours in animals, particularly primates.[41] De Waal's work challenges the assumption that moral

[40] Deane-Drummond, 163.

[41] Ernst Conradie, "Do Only Humans Sin? In Conversation with Frans de Waal," in *Issues in Science and Theology: Are We Special?: Human Uniqueness in Science and Theology*,

behaviour is exclusive to humans and highlights behaviours in animals that resemble empathy and fairness. Conradie, however, draws a crucial theological distinction: while certain animal behaviours may appear morally significant, they do not equate to true moral agency or the capacity for sin. In Conradie's view, sin involves intentionality, moral agency, and, fundamentally, a relationship with God—elements that he considers uniquely human. He uses de Waal's findings as a way to explore the boundary between biological and theological understandings of morality, affirming that sin, as a theological category, pertains to human accountability before God. Conradie argues that, although animals may display behaviours resembling morality, they lack the rational and spiritual faculties required for true moral responsibility. His engagement with de Waal's research reinforces his perspective that sin is a distinctively human phenomenon, grounded in the capacity for a relationship with God and shaped by human intentionality and agency.

While I find Conradie's theological distinction compelling, particularly in maintaining sin as a uniquely human category, my own approach diverges in emphasis. Rather than framing animals exclusively in relation to what they lack (namely, moral agency), I am interested in exploring what they possess: a creaturely capacity to participate in God's life through the Spirit. Even if animals cannot sin, this does not preclude them from being spiritually significant beings who stand in relationship to God in their own way. Extending pneumatological reflection to the animal world allows us to imagine animals as recipients of the Spirit's life-giving presence, not as moral agents but as fellow participants in creation's ongoing story, broadening the scope of God's redemptive purposes without collapsing the theological category of sin into biological behaviour.

In the following section, I will consider various positions regarding animals as non-moral agents. While they are impacted by the consequences of human sin, animals do not bear moral responsibility for the corruption in creation. The purpose of this section is to emphasise the place of animals within the created order: they are bound by natural limits but are exempt from the moral ruptures that stem from human rebellion.

The theological distinction between human and non-human creatures finds its roots in biblical narratives, where animals are often depicted as

ed. Michael Fuller et al., Issues in Science and Religion: Publications of the European Society for the Study of Science and Theology (Springer, 2017), 124–32.

part of a harmonious creation that operates within God-given limits. Old and New Testament texts, while addressing the goodness of creation, also highlight humanity's unique capacity for moral agency and the capacity for rebellion against God. The biblical account seems to distinguish between the moral consequences of the Fall on humans and the absence of such a moral shift in the natural world, including animals.[42] YHWH's speeches in the book of Job, for instance, celebrate the wildness of creation, including predatory behaviour, as a testament to the Creator's design rather than as evidence of moral corruption.[43] Animals, operating within their natural instincts, do not possess the moral capacity for sin or virtue; thus, their predatory behaviour is not framed as moral or immoral but as part of the natural order.[44] Therefore, the realism presented in Job and these early theological reflections suggests that the natural world, with all its complexities and apparent harshness, reflects God's creation rather than a fallen, morally corrupted state.[45]

While non-human animals are similarly impacted by the consequences of sin and can participate in a covenantal relationship with God (e.g., Gen 9:10), they are not morally responsible for the corruption or evil present

[42] Non-human animals are depicted in the Old Testament as possessing a form of accountability, yet this accountability is distinct from the moral culpability associated with human sin. Certain passages (e.g., Gen 3:21; 4:4; 4:7) suggest that animals may serve as atoning sacrifices for human transgressions, which implies that they are innocent of sin in the moral sense. Their status as non-moral agents highlights their inability to bear personal guilt, distinguishing them fundamentally from human beings who are morally accountable before God. This distinction emphasises the theological significance of animals' innocence in sacrificial contexts, underscoring their unique role within the biblical framework of atonement.

[43] See Job 38:39–41; 39:27–30.

[44] In classical Jewish thought, death is sometimes portrayed as a divine punishment or as a consequence of sin, while at other times it is depicted as the natural fulfilment of a complete life ordained by God. The notion that all forms of death in nature are inherently cruel, vicious, sinful, or evil does not originate in ancient Jewish tradition. Osborne, referencing Brueggemann, points out that the Hebrew Bible does not suggest a mechanistic link between sin and death. Osborn, *Death before the Fall: Biblical Literalism and the Problem of Animal Suffering*, 34. Augustine, also echoed this sentiment, viewing natural phenomena, including predation and natural disasters, as integral parts of God's glorious world. Augustine's thoughts will be explored further in Sect. 3.3.2.2.1.

[45] J. Richard Middleton, "Reading Genesis 3 Attentive to Human Evolution: Beyond Concordism and Non-Overlapping Magisteria," in *Evolution and the Fall*, ed. William T. Cavanaugh and James K. A. Smith (CHI: Eerdmans, 2017), 86, http://ebookcentral.proquest.com/lib/harvest/detail.action?docID=4865154.

in the world. Their actions do not stem from moral deliberation, and thus they cannot be considered agents of sin. This is crucial in distinguishing them from humans, who bear moral responsibility for their actions and are accountable for the introduction of sin and corruption into creation.

In the New Testament, non-human animals are depicted as part of the broader redemption from the corruption resulting from human sin (Rom 8:18–22). The suffering and groaning of creation, of which animals are a part, is portrayed as a direct consequence of human transgression, rather than any moral failing on the part of non-human creatures. Their participation in the "bondage of corruption" arises not from their own actions but as an effect of humanity's moral failings. When I speak of the redemption of animals as being tied to that of humanity, I do not mean that animals are morally culpable or in need of atonement. Rather, their liberation is bound up with the eschatological renewal of creation that accompanies the redemption of humanity. Paul's vision in Romans 8:19–22 depicts the whole created order (animals included) yearning for the "revealing of the children of God," when creation itself will be set free from decay. In this way, the restoration of animals is woven into God's redemptive plan for humanity, not as an afterthought but as an integral part of creation's ultimate renewal.

This does not imply that non-human animals are only redeemed if individual humans are redeemed, but rather that the ultimate redemption of the cosmos, which includes animals, is part of the same overarching divine plan that brings redemption to humanity. In this vision, human sin has led to the corruption of the whole created order, and similarly, the redemption of humanity will coincide with the restoration of creation as a whole. Thus, animals are not necessarily dependent on the salvation of individual human souls; rather, they are participants in the cosmic renewal that accompanies the fulfilment of God's redemptive purposes for all of creation.

Therefore, their redemption is tied to the redemption of humanity, which ultimately leads to the liberation of all creation.[46] The language

[46] In this view, the redemption of non-human animals is a component of the broader divine work of restoring the entirety of creation, bringing it into harmony with God's will, where the consequences of sin—including suffering, death, and corruption—are ultimately overcome. This indicates that the redemption of creation, including animals, is an integral part of the consummation of God's kingdom, not a secondary or conditional outcome based solely on the fate of humans.

of creation (κτίσις) in this context points to more than just liberation from suffering—it envisions creation entering into the "glorious freedom" that follows the redemption of human beings.[47] In my view, non-human animals should be understood as non-moral agents—affected by sin's consequences but not responsible for the corruption that afflicts creation.

In *The City of God*, Augustine offers a nuanced consideration of the behaviour of predatory animals within the context of divine providence and natural order.[48] He acknowledges that creatures often deemed savage—such as lions, tigers, and raptors—exhibit behaviours of care and mutuality within their own species. For instance, Augustine notes how a tigress sets aside her ferocity to tenderly nurture her cubs, or how birds of prey engage in cooperative nesting and rearing practices. Such observations point to an inherent structure of order and peace, even amidst predation, which Augustine attributes to the wisdom of divine design. This suggests that the seemingly violent and predatory aspects of nature are not aberrations but are integrated within a purposeful and balanced creation. Augustine's reflections thus challenge simplistic moral assessments of predation as inherently evil, instead inviting recognition of the complex interdependence and equilibrium characteristic of the natural world as an expression of God's providence.

In his reflections on the fall, Dietrich Bonhoeffer provides a clear theological distinction between human and non-human creatures, particularly in terms of moral agency and creatureliness. Bonhoeffer stresses that, through the fall, humanity attempts to transcend its created limits by seeking to "become like God." This rebellion against creatureliness, characterised by the rejection of dependence on God and a desire

[47] The term "ktisis" (κτίσις) is used in Romans 8:19–22. This passage discusses the subjection of creation to futility and its longing for liberation, aligning with the idea of non-human animals being subjected to the consequences of human sin. Romans 8:19–22 (NRSV): "For the creation (ktisis) waits with eager longing for the revealing of the children of God; for the creation (ktisis) was subjected to futility, not of its own will but by the will of the one who subjected it, in hope that the creation (ktisis) itself will be set free from its bondage to decay and will obtain the freedom of the glory of the children of God." This passage is central to the discussion of non-human animals as non-moral agents, as it stresses that creation, including animals, is subject to suffering and decay not by its own fault, but as a consequence of human sin. See David G Horrell, Cherryl Hunt, and Christopher Southgate, *Greening Paul: Rereading the Apostle in a Time of Ecological Crisis.* (Waco, TX: Baylor University Press, 2010), 73–75.

[48] Mary T. Clark, *Augustine*, Outstanding Christian Thinkers (London: Geoffrey Chapman, 1994). *The City of God*, Book XIX, Chapter 12.

for autonomy, is a uniquely human condition. Unlike animals, humans possess the capacity for moral agency, self-consciousness, and the ability to make deliberate choices that lead to disobedience and separation from God.[49] In this theological framework, the fall is not merely a modification of human nature but a fundamental rupture that transforms humanity's very essence. Bonhoeffer suggests that, through the fall, humans lose their original creatureliness, becoming beings who can no longer recognise their dependent status before God. This transformation is specific to humans, as animals—lacking moral agency and the capacity for conscious rebellion—remain within their God-given limits as creatures. Therefore, while humans seek to transcend their creatureliness, animals continue in their natural state of dependence upon God, unaffected by the moral and existential rupture that defines human sin.

This distinction holds weighty implications for discussions of animal moral agency. According to Bonhoeffer's theology, animals, unlike humans, do not participate in the moral rebellion of the fall. They remain part of the created order, operating according to their natural instincts without the capacity for moral responsibility. Thus, Bonhoeffer's view aligns with a broader theological position that animals, while integral to creation, do not possess the same moral agency as humans and are not implicated in the fall in the same way.

Recent theological discussions have expanded upon traditional perspectives by critically examining the moral and spiritual status of animals within creation.

In his theological vision, Moltmann presents a cosmology wherein God is profoundly committed to creation, sustaining, and nurturing all living beings within an interconnected community. His pneumatology proposes a God who is both transcendent—granting creation its distinctness and autonomy—and immanent, intimately involved in preserving the vitality of all creaturely life.[50] This dual aspect creates a framework of divine patience and forbearance, allowing creation the time and space necessary for growth, change, and eventual restoration. God's patience, marked by a willingness to bear suffering, reflects a hopeful anticipation

[49] Dietrich Bonhoeffer, *Creation and the Fall: A Theological Interpretation of Genesis 1–3*, trans. John C Fletcher (London: SCM Press, 1959), 72–73.

[50] Moltmann, *The Spirit of Life*, 42.

that all creatures will ultimately turn back towards God's Kingdom.[51] Within this cosmic scope of redemption, Moltmann affords reverence to all life forms, emphasising that Christ's incarnation positions him as the ultimate participant in the suffering inherent in evolutionary processes. Through his resurrection, Christ gathers all suffering creatures, inviting them into the life of the Trinity.[52] I find Moltmann's framing theologically compelling as he refuses to romanticise death or treat it as a neutral necessity. His statement that "the death of the living is neither due to sin nor is it natural" underscores this stance.[53] For Moltmann, death is neither a moral fault nor a harmonious element of creation; it is a tragic and alien feature within the created order—something God mourns and ultimately works to overcome.[54] In this framework, animals are not morally culpable beings; they do not bear the weight of sin in the human sense. Instead, they participate in a shared cosmic longing for the future world and eternal life. Thus, Moltmann's theology presents animals as innocent participants in creation, affected by mortality but free from moral guilt. Ultimately, they too are included in God's eschatological vision of the "deification of the cosmos," wherein all creatures will partake fully in divine life, unblemished by the taint of sin.

Andrew Linzey's theological reflections on animals as moral agents reveal a nuanced understanding of animal suffering and predation within the framework of a fallen creation. Linzey resists attributing moral agency to animals, suggesting that behaviours such as predation and parasitism, while indicative of suffering and violence, do not constitute moral evil in the same sense as human sin.[55] Rather, he argues, these aspects of animal life reflect an "incomplete" creation impacted by a cosmic fall

[51] Jürgen Moltmann, *The Way of Jesus Christ: Christologyin Messianic Dimensions*, trans. Margaret Kohl (MN: Fortress Press, 1993), 290.

[52] Moltmann, 153; Jürgen Moltmann, *In The End- The Beginning: The Life of Hope*, trans. Margaret Kohl (MN: Fortress Press, 2004), 70.

[53] Jürgen Moltmann, *The Coming of God: Christian Eschatology*, trans. Margaret Kohl (MN: Fortress Press, 1996), 92.

[54] This resonates strongly with the biblical narrative Koyama highlights: history is not destined to endlessly repeat cycles of suffering and death but is moving toward a divine victory that will liberate creation from death's reign. See Kosuke Koyama, *Water Buffalo Theology* (Maryknoll: Orbis Books, 1999), 85. I thank Frank Macchia for bringing this valuable resource to my attention.

[55] Andrew Linzey, *Animal Gospel* (Louisville, KY: Westminster John Knox Press, 1998), 28–29. See also Linzey, *Why Animal Suffering Matters*, 22–25.

that disrupts the intended harmony of the natural world. For Linzey, this cosmic disorder profoundly affects sentient beings, yet it does not imply moral culpability on the part of animals. Animals, he argues, operate according to their natural instincts and are bound within the structures of the created order without the capacity for intentional wrongdoing. Humans alone, in Linzey's view, bear moral responsibility for sin and suffering, as they possess the unique capacity for deliberate moral choice and thus carry the ethical weight of creation's disorder.[56]

Linzey's approach raises crucial theological questions concerning divine providence and the eschatological hope for redemption. He is careful to affirm that while predation and suffering are woven into the fabric of the current natural order, they are not expressions of divine intent.[57] Instead, Linzey suggests that these phenomena await the transformative work of redemption, challenging any theological stance that would accept the violent aspects of nature as morally neutral or divinely sanctioned. Linzey's position reflects a deep ethical commitment to animal welfare and to a vision of God who does not abandon creation to cycles of suffering. His insistence that animals are recipients of redemption rather than morally culpable actors aligns closely with my own view that animals, while affected by the consequences of sin, are not themselves sinners. Linzey calls human beings, as moral agents, to bear witness against the suffering embedded in nature, joining in the divine mission to liberate creation from its fallen state. I find this to be a powerful theological challenge: it frames human ethical responsibility not simply in terms of stewardship but as active participation in God's redemptive work.

Linzey's reflections contribute a distinctive voice to theological discussions on animal suffering. By positioning animals as innocent yet integral participants in eschatological hope, he offers a framework that both affirms their value and implicates humanity in the task of anticipating and embodying the redemption still to come.

Ernst Conradie approaches the concept of sin not in traditional theological terms, but as a form of "social diagnostics."[58] He suggests that Christian discourse on sin offers an interpretative lens for diagnosing what

[56] Linzey, *Animal Gospel*, 32.

[57] Ryan Patrick McLaughlin, "Andrew Linzey (1952-): Animal Theology," in *Animal Theologians*, ed. Andrew Linzey and Clair Linzey (NY: Oxford University Press, 2023), 429.

[58] Conradie, *Redeeming Sin?*, xxv.

has gone wrong in the world, particularly in light of ecological destruction. Sin, in this approach, is less about individual moral failures and more about underlying structural problems—forms of alienation from God that manifest as ecological harm, inequality, and social injustice. Sin-talk, according to Conradie, provides a way to understand these deeper issues and their pervasive impact, which are best understood through the lens of a broken relationship with the triune God. Conradie recognises that animals may exhibit behaviours that resemble moral actions—such as cooperation, nurturing, or aggression—but he does not consider them moral agents in the same way humans are. While studies in animal ethology suggest forms of "protomorality" (e.g., deliberation, restraint, and cooperation), Conradie is careful to distinguish these behaviours from human moral agency, which involves higher cognitive functions such as abstraction, symbolic expression, and emotions like shame or guilt.[59] Conradie acknowledges that animals' behaviours can be assessed in terms of value but stresses that not all value judgements are moral judgements. For him, only beings capable of making moral choices—namely, humans—can be considered moral agents. This means that while animals play significant roles in ecological systems and exhibit complex social behaviours, they are not responsible for the moral ruptures, such as sin, that Conradie associates with human actions. Conradie also hints that there may be some graded understanding of agency, suggesting that some animals could be considered closer to moral agents than human infants or those who lack moral agency due to disability. However, this remains speculative and does not imply that animals are moral agents in the same sense as fully developed human beings.[60] Thus, in his framework, animals are affected by the broader consequences of human sin and natural processes but are not morally culpable or capable of moral responsibility. This reinforces the idea that animals exist within the created order as beings bound by their natural instincts and limitations.

In the preceding sections, I have examined a range of perspectives on animals' non-moral agency, particularly as it relates to the problem of natural suffering. I now turn to consider the Spirit's presence within the intricate and often brutal realities of predation. Whether one interprets creation as fallen or unfallen, or debates the moral capacities of animals,

[59] Conradie, 37.

[60] Conradie, 214.

the Spirit's work is not bound by these theological or philosophical categories. The Spirit moves within and through these realities, accompanying creation in all its complexity. What follows explores how the Spirit is dynamically present in the lives of animals—not apart from the violent and painful contexts of predation and survival, but precisely in their midst.

Pneumatological Considerations

This section will focus specifically on the pneumatological aspects of predation, exploring how the Holy Spirit engages with this natural process, particularly in relation to animals. I will not attempt to resolve the problem of predation and suffering. Instead I will highlight several key insights from a pneumatological perspective that emphasise the Spirit's role in sustaining, guiding, and transforming creaturely life in their evolutionary reality. These reflections will consider the implications for animals, whose experiences of predation and survival are integral to the Spirit's creative and redemptive work within the natural order.

Within the animal kingdom, predation serves as a fundamental process for maintaining ecological balance and promoting biodiversity. From a pneumatological perspective, the Holy Spirit can be understood as actively sustaining animal life, not only through creation but also through the cycles of death that are intrinsic to predatory relationships. Denis Edwards highlights that death, including predation, is essential for the unfolding of life and the evolutionary development of species. The Spirit, as the Breath of Life, is present in this cycle, ensuring that predation does not simply signify loss but is an integral part of the web of life. The death of one animal provides sustenance for another, contributing to the flourishing of ecosystems as a whole. In this way, the Spirit works through the natural predatory behaviours of animals, transforming what may appear as brutal necessity into a deeper ecological interdependence where life feeds life. Whether predation is viewed as part of a fallen creation or as inherent to the natural world, the Spirit's presence within these processes affirms that animal suffering and death are not devoid of purpose. Instead, they are held within the broader redemptive work of creation, where the Spirit sustains, renews, and ultimately guides creation—including non-human animals—towards fullness and flourishing.

Predation can be seen as part of the unfolding complexity of creation, which is both beautiful and brutal. From a pneumatological standpoint, the Spirit's role in creation is to guide creation through this complexity, ensuring that even the painful realities of predation are part of the journey towards fulfilment and transformation. The Spirit is actively present in these cycles, not eliminating predation but working through it to bring about new life and to draw creation towards its ultimate eschatological renewal. Predation, in this sense, is one part of a larger, Spirit-guided process towards the reconciliation of all things.

Amos Yong suggests that natural processes, including those perceived as evil, are part of an unfolding, Spirit-guided creative act. This pneumatological lens allows for a reinterpretation of natural suffering as integral to the interconnectedness and interrelatedness of creation, where the Spirit works through complexities and apparent randomness to sustain life and foster transformation.[61] Yong's approach highlights that the Spirit is not merely a presence of order and harmony but engages actively with creation's disorder, guiding it towards greater complexity and relational depth. A key element in Yong's argument is the *ruach Elohim* (Spirit of God) in Genesis 1, traditionally understood as "hovering" over the chaos at the beginning of creation. Rather than viewing this initial state as an empty void, Yong interprets it as a state of plenitude and potential, where differentiation and complexity can emerge. Through the lens of emergence theory, Yong suggests that creation itself evolves through Spirit-driven processes, marked by dynamic, non-linear relationality between God, creation, and creatures. In this sense, creation is not a static, harmonious entity but a continuously unfolding reality—*creatio continua*—in which natural suffering, such as predation and pain, is woven into the Spirit's sustaining and transforming work.

Yong's view proposes that natural suffering is not external to the Spirit's work but rather a facet of creation's evolving complexity and synergy. Such a perspective challenges traditional interpretations of natural suffering by positing that even disorder and pain are part of a larger movement towards reconciliation. However, this approach also raises significant theological questions. Does reframing natural suffering as a component of the Spirit's creative process risk diminishing the real experience of suffering? Might it challenge classical conceptions of divine goodness,

[61] Yong, *The Cosmic Breath: Spirit and Nature in the Christianity-Buddhism-Science Trialogue*, 58–70. See also Amos Yong, *The Spirit of Creation: Modern Science and Divine Action in the Pentecostal- Charismatic Imagination* (Grand Rapids, MI: Eerdmans, 2011), 164.

especially for those who interpret natural suffering as incompatible with a loving God?

Yong's pneumatological framework calls for a careful balance: recognising that the Spirit is at work even amidst creation's darkest elements, yet without overlooking the pain and suffering experienced by creatures. By situating natural suffering within the broader narrative of creation's unfolding complexity, Yong presents a theological vision that holds both beauty and brokenness, interdependence and discord, as intrinsic to creation's Spirit-guided journey towards a reconciled state. This perspective aligns with an eschatological hope in which the Spirit, working within the fabric of creation itself, guides all things towards a final restoration that transcends the limitations and suffering present within the current natural order.

Drawing on Augustine, James K.A. Smith argues that miracles should not be understood as violations of natural laws but as intensified expressions of the Spirit's ongoing creative activity.[62] In this sense, even

[62] James K. A. Smith, "Is the Universe Open for Surprise? Pentecostal Ontology and the Spirit of Naturalism," *Zygon: Journal of Religion & Science* 43, no. 4 (December 2008): 892. Smith critiques a common Pentecostal and charismatic tendency to over-emphasise the miraculous aspects of the Spirit's work, which often leads to an underappreciation of the Spirit's presence in the mundane and ordinary processes of creation. Smith terms this as a "hyper-spiritual" focus, wherein believers might expect divine action only through supernatural interventions, neglecting the Spirit's continuous engagement with the natural world, including processes such as predation. For Smith, the Spirit's dynamic presence is not restricted to miraculous events but is woven through the fabric of everyday life, including the natural cycles of death and renewal, offering a more holistic vision of divine activity that encompasses both the extraordinary and the ordinary. However, in his broader theological reflections, Smith issues a caution against "ontologising" the Fall, whereby sin and brokenness become inscribed into the very fabric of creation. He warns that such views risk embedding sin as an inherent feature of the natural order, thereby blurring the line between creation's original goodness and the corruption introduced by human rebellion. This concern is particularly significant for theological anthropology, where Smith resists any suggestion that brokenness or violence is "natural" or divinely intended. Instead, he underscores that grace should not oppose or nullify nature but rather restore or perfect it, in line with the broader Christian tradition's resistance to any Gnostic dualism that might view redemption as antithetical to the created order. James K. A. Smith, "What Stands on the Fall? A Philosophical Exploration," in *Evolution and the Fall*, ed. William T. Cavanaugh and James K. A. Smith (Chicago, IL: Eerdmans, 2017), 65.

These two perspectives may appear in tension, but they reveal a complex theological vision. Smith's reading of predation as part of the Spirit's work within a fallen creation does not imply that he views predation as ultimately good or eschatologically ideal. Rather, his position can be understood as a pneumatology that acknowledges the Spirit's sustaining

predation can be seen as part of the Spirit's ordinary yet dynamic work in creation. The Spirit does not eliminate predation but operates through it to bring forth life, complexity, and the flourishing of ecosystems. Like Yong, Smith affirms that the Spirit's role in creation is not merely to maintain order but to sustain the evolutionary processes that lead to transformation and fulfilment.

Both Yong and Smith highlight the Spirit's constant, yet often overlooked, engagement with creation. Whether through animal predation or other forms of natural processes, the Spirit guides creation towards its eschatological renewal. In this framework, predation is not viewed as a manifestation of divine absence or neglect but as part of the broader narrative of the Spirit's involvement in non-human creation's journey towards ultimate reconciliation and peace.

Before turning to a pneumatological interpretation of predation as a life-sustaining process, it is important to pause and address the deeper theological paradox that underlies these reflections. At the heart of this discussion lies the contested status of death itself: is death a natural and necessary part of creation's flourishing, an evil to be overcome, or in some sense both? How we answer this question shapes not only our theological interpretation of predation but also how we understand the Spirit's role within these realities.

A central theological tension emerges when predation is situated within the broader question of death's status in creation. Death is, on the one hand, essential to the functioning of the natural world as we know it: evolutionary processes depend on predation and decay to sustain biodiversity and ecological balance. On the other hand, Christian eschatology

and redemptive work within the realities of a broken world while holding that these realities point beyond themselves toward a reconciled and peaceable creation. The Spirit operates within the evolutionary processes of the current order, guiding life's resilience and interdependence, even as it moves creation toward an eschatological horizon where such violence is no longer necessary. In this sense, Smith's work presents a pneumatological framework that is both attentive to the Spirit's redemptive role in sustaining the complexities of a fallen creation and anticipatory of an eschatological peace. His theology maintains the distinction between creation's inherent goodness and the distortions introduced by sin, aligning the Spirit's work with a vision of cosmic redemption that ultimately transcends the predatory structures of the current order. Through this dual focus, Smith's reflections contribute a dynamic yet critical voice within contemporary discussions on the Spirit's involvement in natural processes, offering a vision of creation that, while marked by suffering, is ultimately oriented toward restoration and fulfilment in the life-giving presence of God.

proclaims death's ultimate defeat, casting it as an enemy to be overcome (1 Cor 15:26). This paradox—death as both indispensable to life now and yet destined for eradication—demands theological interrogation rather than quiet acceptance.

Andrew Linzey's insistence that the violent cycles of animal life cannot be equated with God's true design sharpens this tension. The Edenic vision gestures towards a peaceable creation in which predation does not structure inter-creaturely relations, while the evolutionary narrative portrays predation as a primary engine of life. How, then, can predation be understood theologically in light of death's eschatological defeat?

A pneumatological perspective offers a way to hold these poles together without collapsing one into the other. The Spirit is present within the current order, sustaining life through the very cycles of death and predation that characterise evolutionary history. In this sense, death is not external to the Spirit's creative work but functions as part of the provisional means by which creation flourishes in the present age. Yet the same Spirit is also the agent of eschatological renewal, the one who will ultimately liberate creation from its bondage to death and bring it into the fullness of divine life (Rom 8:21).

In this light, predation occupies an ambiguous theological status. It is both an instrument through which the Spirit presently sustains life and a manifestation of the mortality that will one day be overcome. Rather than romanticising predation as simply "natural" or rejecting it outright as evil, this framework holds it in eschatological tension: it is provisionally permitted, Spirit-sustained, yet ultimately subject to transformation and defeat.

This pneumatological framing refuses to sacralise the current natural order while also affirming that God is not absent from it. The Spirit accompanies creation through the cycles of life and death, even as these cycles are destined to give way to a new reality in which death and predation no longer define creaturely existence. In this way, death is treated neither as a mere neutral feature of nature nor simply as an evil intrusion, but as a contested and provisional reality within the drama of creation and redemption—a reality the Spirit both sustains and will finally transcend.

Having acknowledged this tension, it becomes possible to consider more closely how the Spirit is presently at work within these provisional realities. Predation, though bound up with death and suffering, also participates in the Spirit's sustaining and creative activity. A pneumatological lens allows us to examine not only how predation will be ultimately overcome, but also how it currently functions within the unfolding of creation's story. Denis Edwards observes that "without death, there would be no series of generations, and without the series of generations, there would be no evolution", drawing attention to the indispensable role that death (including predation) plays in the unfolding of life.[63] From a pneumatological perspective, the Spirit animates and sustains life not only in its flourishing but also through the cycles of death and predation that structure ecological systems.

Whether natural suffering is understood as the consequence of a fallen creation or an inherent part of the natural world, the Spirit is not absent but remains dynamically present within these realities. The Spirit's work moves through predation, renewing and sustaining creation by enabling its continuous development.[64] Predation, rather than being viewed solely as a violent manifestation of natural suffering, can be understood as part of the Spirit's ongoing creative activity (irrespective of its fallen/unfallen state)—ensuring the flourishing and resilience of life through evolution and the interdependence of ecosystems. The Spirit works within these predatory processes to bring forth life and to maintain the complex balance of creation.

While predation may seem antithetical to notions of divine goodness, a pneumatological approach could frame it as a redemptive opportunity. The Spirit does not simply tolerate predation; the Spirit inhabits these relationships, infusing them with purpose and directing them towards the flourishing of creation. What seems at first to be a brutal necessity is reoriented within the broader divine economy as the Spirit works to redeem the pain and loss inherent in predation, drawing it into the larger narrative of creation's renewal.

The Holy Spirit, present in every dimension of creation, also participates in the groaning and suffering of animals, echoing the groans of creation described in Romans 8. Elizabeth Johnson speaks to the compassionate character of the Spirit, who, in the midst of both agony and delight, does not remain distant but is deeply engaged with the suffering of all creatures. Johnson's framing envisions the Spirit's nature as one of

[63] Edwards, *Breath of Life*, 137. Elizabeth Johnson also makes this point in conversation with John Haught, see Johnson, *Ask the Beasts*, 189.

[64] Johnson states, "The Spirit's action does not supplant that of creatures, but works cooperatively in and through created action, random, ordered or free." Johnson, *Women, Earth, and Creator Spirit*, 58.

compassion, resistance, reconciliation, and accompaniment.[65] The Spirit does not merely observe animal suffering but is present within it—sympathising, comforting, and liberating (irrespective of the moral status of a non-human animal).

Pneumatologically, pain can be seen not merely as an isolated negative experience but as part of the Spirit's transformative journey that leads to peace and fulfilment. This vision is grounded in the idea that the Holy Spirit works within the experiences of creation—including suffering—to bring about growth, transformation, and an eventual state of harmony. Pain becomes a conduit, or a vehicle, that propels beings from a place of fragmentation and imperfection towards a state of wholeness. The idea of pain as a vehicle towards peace resonates with the narrative of Christ's passion and crucifixion. The cross represents the pinnacle of suffering—brutal violence inflicted upon the innocent—and yet, through the power of the Holy Spirit, it becomes the very means through which reconciliation, healing, and ultimate peace are achieved. Through Christ, death is conquered. Similarly, the suffering experienced by non-human creatures is not meaningless; it is caught up in the Spirit's redemptive activity, which does not justify pain but refuses to leave it without purpose. In this way, the Spirit works through the experiences of suffering in creation, transfiguring them into integral moments within the larger drama of cosmic redemption.

These pneumatological considerations do not fully resolve the issue of natural suffering, but they do highlight the Spirit's active involvement in the lives of non-human animals. The Spirit is engaged in the harsh and violent realities of evolution, including predation and suffering, working in the midst of these processes rather than avoiding them. The Spirit is present with animals in their past, present, and future, sustaining them through evolutionary challenges and pointing towards a future hope of renewal and reconciliation. This framework reflects a vision where the Spirit remains with creation, even amid suffering and death.

[65] Johnson, 59.

Conclusion

In this chapter, I have explored the pneumatological dimensions of natural suffering, with particular attention to predation and its significance for non-human animals. While these reflections do not resolve the enduring theological tensions surrounding natural suffering, they offer a way of perceiving the Spirit's presence not at the margins, but at the very heart of creation's unfolding drama. The Holy Spirit is shown to be dynamically involved in the evolutionary history of life—present in moments of flourishing as well as in the violent and painful realities of predation and death. As the breath that animates and sustains, the Spirit neither eradicates these realities nor stands apart from them, but works through the very processes that sustain ecological life, holding together a creation marked by both beauty and brutality.

This pneumatological vision refuses simplistic categorisations of death and predation as wholly natural or wholly evil. Instead, it frames them as contested and provisional realities within God's redemptive economy—realities that the Spirit both upholds and will ultimately transcend. The Spirit's work is thus both sustaining and eschatological: accompanying creation in its present cycles while drawing all creatures towards their ultimate renewal.

Further theological reflection could fruitfully build upon this foundation, tracing how the Spirit's presence amid non-human suffering intersects with wider ecological and soteriological themes. Such inquiry may deepen our understanding of how divine action and creaturely experience intertwine in a world that is simultaneously wounded and wondrous. By attending to the Spirit's work within the fraught realities of the natural world, theology may better articulate a vision of creation in which even its most painful dimensions are neither ignored nor romanticised, but are held within the Spirit's patient, redemptive embrace.

CHAPTER 4

Soteriology

In the previous chapter, I raised the question of whether animals can be regarded as moral agents, or whether they stand outside that category altogether. If they are moral agents, capable of bearing responsibility for their actions, then the scope of the cross might be understood as extending even to non-human creatures. Yet if they are not, animals nevertheless remain enmeshed in a creation marked by the fall. Their suffering, as well as the degradation of the environments they inhabit, bears witness to a world fractured and in need of healing. Redemption, then, is not required because animals themselves are guilty of sin, but because, along with all creation, they endure the disorder and decay introduced by it. Christ's redemptive work, therefore, cannot be confined to the problem of human transgression alone; it embraces the renewal of the whole created order. Animals are included in the broader scope of cosmic redemption, whose ultimate goal is the restoration of all creation to its intended harmony and flourishing. Their redemption is part of the transformation of creation, through the work of Christ and the Spirit, into the fullness of life that God intends. I reinforce that I do not consider animals as moral agents and that they are therefore not "fallen" in the human sense. Scripture demonstrates that sanctification is not limited to moral beings (Gen 2:3; Exod 29:43; Lev 16:19). God sanctifies time, places, vessels, and land, setting them apart for divine purposes. Likewise,

D. Rizzo, *Animal Pneumatology*, The Palgrave Macmillan Animal Ethics Series, https://doi.org/10.1007/978-3-032-18723-9_4

the Spirit may consecrate creatures for praise and service without implying moral culpability.

Animal theology has long centred on Christological themes, particularly focusing on soteriology and incarnation.[1] Scholars like Andrew Linzey, David Clough, and Richard Bauckham have shaped this conversation by examining the redemptive work of Christ and the incarnation's implications for human-animal relationships. Linzey highlights the significance of the incarnation for all living beings, contending that the *Logos* identifies not only with humanity but with every creature of flesh and blood. Clough expands the scope further, envisioning a cosmic Christology that embraces the shared life of humans and animals alike. Bauckham explores the ecological reach of the incarnation, seeing in the Word made flesh a profound connection to the physical nature shared by all living things. Together, these voices have deepened our understanding of how Christ's salvific work extends to the whole of creation, offering fresh perspectives on the place of animals within God's redemptive plan.

While honouring the significant contributions of these Christological insights, this chapter seeks to extend the conversation further by venturing into the realm of Pneumatology.[2] In engaging with both Christological and Pneumatological dimensions, I seek to demonstrate that the Spirit's life-giving and grace-filled presence is not only at work in creation but also deeply involved in the redemptive trajectory set in motion by the incarnation. The Spirit's baptising work, which extends to "all flesh," highlights a radical connection between creation, incarnation, and the Spirit's continuous role in sustaining and transforming the entirety of the created order—human and non-human alike. By incorporating a Pneumatological lens, this chapter aims to deepen the understanding of how the Spirit actively participates in and affirms the intrinsic value and role of all creatures. The aim is not to diminish the importance of existing Christological discussions but to enrich them. By threading Pneumatology into

[1] Part of this chapter has been published in Daniela Rizzo, "The Word Made Flesh and the Spirit's Breath: Bridging Christology and Pneumatology in Animal Theology," *Religions* 15, no. 11 (2024): 1355, https://doi.org/10.3390/rel15111355.

[2] This chapter responds to Denis Edwards' invitation to develop a systematic theology of deep incarnation by tracing the intimate bond between the Spirit's life-giving work and the incarnation itself. As Edwards observes, "A systematic theology of deep incarnation will need to show how the Spirit who is giver of life in creation and in grace is radically connected to the incarnation itself." Denis Edwards, *Deep Incarnation: God's Redemptive Suffering with Creatures* (Maryknoll, NY: Orbis Books, 2019), 107.

the fabric of soteriology and incarnation, this chapter offers a new lens through which to view the role of animals within the divine story of salvation and the "all flesh" baptism brought forth by the Spirit.

Viewed through a pneumatological lens, a wider horizon comes into focus—one in which the Spirit's baptising work encompasses the whole community of life. Following this trajectory, the chapter seeks to cultivate a more expansive theological vision, one that recognises animals as active participants in the divine drama of redemption and extends animal theology into the rich, life-giving currents of pneumatology.

Theological Insights of Soteriology and Non-Human Creatures

In Scripture, God is revealed not only as the Creator but also as the compassionate Redeemer and Saviour of the world—a story in which animals, too, find their place. Existing animal theologies have rightly recognised animals as recipients of salvation, a theme to which I now turn more fully.[3] Building on this foundation, Ryan Patrick McLaughlin's engagement with John Wesley's theology offers a particularly fruitful lens for considering the wider scope of redemption. By revisiting Wesley's often-overlooked sermons, "The General Deliverance" (1781) and "The New Creation" (1785), McLaughlin uncovers a nuanced and expansive vision of salvation that embraces not only humanity but the whole of creation, including non-human animals. Wesley's understanding of

[3] This section does not aim to provide an exhaustive literature review on animal soteriology. Instead, it will highlight and engage with some of the key contributions that frequently appear in discussions within the field of animal theology.

It can be argued that salvation cannot be based merely on incarnation. While this chapter emphasises the integration of Pneumatology with incarnation, the monograph incorporates the six Christological symbols in varying degrees. Incarnation, Pentecost, resurrection, and *parousia* are explicitly addressed, with detailed discussions on Christ's identification with all creation, the Spirit's outpouring on "all flesh," and the eschatological renewal of all creatures. The cross and ascension are more implicitly woven into the narrative, particularly through themes of redemption and the Spirit's ongoing transformative work following Christ's ascension. Together, these symbols form a comprehensive framework for understanding salvation, encompassing both human and non-human creation.

atonement extends beyond human concerns to encompass the often-forgotten participants in God's redemptive drama.[4] Yet, McLaughlin also draws attention to a persistent tension in Wesley's thought: the cruelty evident in nature sits uneasily alongside his theological convictions about God's mercy, justice, and care for all creatures. In wrestling with this dissonance, Wesley turned to Eastern Orthodox notions of sacramentality, envisioning creation as a meeting place between the divine and the creaturely—a shared space in which humans and non-human animals coexist in harmony as part of God's ultimate design. Furthermore, Wesley rejected sharp distinctions between humans and animals regarding immortality, underscoring the radical inclusivity of his soteriology. McLaughlin highlights Wesley's view that humans, endowed with the unique capacity to know, love, and obey God, were entrusted with a sacramental vocation: to sustain harmony with the non-human world. It is the rebellion of humanity, Wesley argued, that shattered this harmony, unleashing cosmic disorder and the suffering of all creatures.

This view of God's redemptive work is presented in prominent biblical passages, which emphasise God's steadfast love and salvific care for all creatures, human and animal alike, as seen clearly in Psalm 36. Prominent biblical passages portray the Creator as a God characterised by loving kindness and fidelity, not merely as the originator of creation but also as the redeemer and saviour of all that has been brought into existence, including animals. This perspective finds explicit affirmation in Psalm 36: "Your steadfast love, O Lord, extends to the heavens, your faithfulness to the clouds. Your righteousness is like the mighty mountains, your judgments are like the great deep; you save humans and animals

[4] McLaughlin examines Wesley's connection of all creaturely suffering and death to the Fall, acknowledging potential scientific challenges to this theological stance. He highlights Wesley's speculative reflections on the extent of loss experienced by nonhuman creatures, proposing a degradation in their understanding, will, and liberty. The acknowledgment of predatory nature in some creatures introduces the concept of "original sin" in the nonhuman realm. Ryan Patrick McLaughlin, "Redemption in John Wesley's The General Deliverance: Theological Hope vs. Biological Reality." (Wesleyan Theological Society, Nampa, ID: Nampa Nazarene University, 2014), 8. Furthermore, McLaughlin notes Wesley's controversial rejection of essential differences between humans and nonhumans, including the contentious issue of immortality. Here, McLaughlin emphasises Wesley's argument that humans, endowed with the capacities for knowing, loving, and obeying God, played a sacramental role in the harmony with the nonhuman creation. The rupture of this harmony is attributed to the rebellion of humans, resulting in cosmic chaos and the widespread suffering of all creatures. McLaughlin, 8.

alike, O Lord."[5] This declaration unambiguously includes animals in the vast scope of God's saving grace, emphasising their integral place in the divine plan of redemption. Examining this psalm, Johnson brings attention to the significance of the verb within the Hebrew Bible, *yasa*, translated as "to save". This verb, recurrent across diverse contexts, is a foundational term denoting salvation, deliverance, and liberation, often attributing these actions to God as the ultimate saviour. Ken Stone notes: "By extending God's salvation to encompass animals alongside humans and utilising such terminology to convey this concept, Psalm 36 … places animals firmly within the redemptive activity of God that is often understood as central to biblical religion."[6] This insight positions animals as active participants in the redemptive narrative, sharing in the divine plan of salvation.

Extending this vision of salvation, Edwards articulates how co-creaturely kinship ultimately reaches its fulfilment in the eschatological resurrection, where all creation is gathered into union with the risen Christ. In this framework, the risen Christ functions as the ecological centre of creation, the one in whom all things find their true coherence and completion in God. Edwards envisions the incarnate Word relating not only to humanity but also to other species and even inanimate nature in a profoundly relational way. The transformative scope of the incarnation encompasses the whole created order: through the crucified Christ's loving self-identification with creation—in both its disharmony and decay as well as its vitality and abundance—creation is drawn into the redemptive movement of God. Bauckham similarly contends that the risen Christ draws the entirety of creation into the eschatological newness of resurrection. In this way, the incarnation is understood not merely as a human-centred event but as salvific for all creatures.[7]

[5] Johnson, "Animals' Praise of God," 265. In response to this inclusive salvation, Johnson notes that animals are depicted as actively seeking and waiting for the Lord (Ps 145:15), actions described using the same verb employed elsewhere when a human expresses hope for God's salvation (Ps 119:166). Within the encompassing circle of salvation, animals place their trust in their Creator, whose loving-kindness extends to care for their lives. This depiction reinforces that animals, too, find refuge and hope in the salvific actions of their Creator.

[6] Stone, *Reading the Hebrew Bible with Animal Studies*, 142.

[7] Edwards, *Deep Incarnation*, 111. Edwards adds, "I think it can be said that God does not forget any creature that God loves and creates, but inscribes it eternally in the divine life." Denis Edwards, "Redemption of Animals in an Incarnational Theology," in

Deep Incarnation

In studies of the incarnation, much has been said about Jesus's Jewishness, maleness, and humanness. However, the theological idea of deep incarnation highlights the significance of Jesus's creatureliness. This section does not aim to dismantle the insights of deep-incarnational theologians, but rather to stand alongside them, offering a contribution in harmony with their work. Far from retreating from the scandal of Jesus's particularity, I suggest that it is precisely in and through Jesus of Nazareth that the horizon widens, making room for a more expansive theological imagination—one that embraces other-than-human animals within the mystery of incarnation.

When considering the incarnation, theologians have traditionally interpreted the term "flesh" to signify Jesus's human nature, taking an anthropocentric position that equates "flesh" specifically with humanity.[8]However, contemporary scholars such as Andrew Linzey, David Clough, Richard Bauckham, Denis Edwards, and Mark Wallace have largely departed from an anthropocentric understanding and have argued for a broader interpretation. They extend the significance of "flesh" beyond humanity and human experience. This idea, more popularly known as "deep incarnation", was originally coined by Niels Gregersen and originated as a concept for comprehending Christology

Creaturely Theology: On God, Humans and Other Animals, ed. Celia Deane-Drummond and David Clough (London: SCM Press, 2009), 96.

[8] Recent literature on ecotheology and theology concerning nonhuman animals has focused on the Gospel of John's statement that the Word became flesh (John 1:14). Although it can be argued that Barth introduces the possibility of extending the Incarnation to nonhuman creation without introducing notions of sinful flesh into the Gospel. For Barth, " 'the Word became flesh' also means 'the Word became time…' It does not remain transcendent over time, it does not merely meet it at that point, but it enters time; nay, it assumes time; nay, it creates time for itself." Karl Barth, *Church Dogmatics*, vol. 1/2 (Edinburgh: T&T Clark, 1956), 50. This perspective challenges the limitation of "flesh" to exclude nonhuman animals. If God encompasses concepts like "time," then it logically follows that the Incarnation extends to include nonhuman animals. Flesh, understood as continually moving through time, undergoes constant change, shedding and replacing cells, suggesting a perpetual state of becoming. The idea of becoming "flesh" doesn't imply a static identity; rather, it indicates a dynamic process. The incorporation of time into the Incarnation emphasises its vulnerability—the Word taking on flesh that ages, resulting in visible changes to one's identity.

within an evolutionary framework.[9] This expanded perspective has not only challenged traditional anthropocentric views but has also influenced the animal theologies of pioneering scholars in this field.[10] In this section, I examine current literature on deep incarnation to explore how Christ's creaturely nature is theologically integrated into the broader context of creation, particularly in relation to non-human creatures.[11]

One of the earliest and most influential attempts to extend the concept of the incarnation beyond the human sphere comes from Andrew Linzey.[12] In his introduction to the edited volume *Animals on the Agenda*, Linzey pointedly declares that "by becoming flesh, the Logos identifies… not only with humanity but also with creatures of flesh and

[9] According to Gregersen, Deep Incarnation signifies the *Logos* entering not only into humanity but also into the entire fabric of biological existence and the natural system. Niels Henrik Gregersen, "The Cross of Christ in an Evolutionary World," *Dialog: A Journal of Theology* 40, no. 3 (September 2001): 192–204, https://doi.org/10.1111/0012-2033.00075.

[10] Matthew Eaton provides a doctrinal genealogy of Deep Incarnation. See Eaton, *Incarnate Earth*, 128.

[11] Surveying the entirety of literature on deep incarnation would be an extensive undertaking within the confines of this chapter. As I delve deeper into this concept, I recommend interested readers to explore the comprehensive scholarship of Matthew Eaton. Matthew Eaton, *Incarnate Earth: Deep Incarnation and the Face of Christ* (London: Routledge, 2023), https://doi.org/10.4324/9781003287346.

[12] In addition to what I cited in the text above, see also Andrew Linzey and Dan Cohn-Sherbok, *After Noah: Animals and the Liberation of Theology* (London: Mowbray, 1997), 85; 103–104.; and Linzey, *Animal Gospel*, 7. While I have highlighted only select examples from the past two decades, a truly comprehensive account would need to engage the extensive corpus of Andrew Linzey's work. Over the course of his career, Linzey has produced a rich and interdisciplinary body of scholarship—ranging from his unpublished dissertation on Karl Barth and nonhuman animals to his seminal *Animal Theology*—which spans theology, biblical studies, ethics, and the social sciences. His contributions are foundational, laying much of the groundwork for the recent flourishing of scholarship on Christian theology, Scripture, and ethics in relation to nonhuman animals. Linzey is among the pioneers who underscored the theological significance of John's prologue and the incarnation for understanding the divine assumption of "flesh." Throughout this chapter, I draw selectively on Linzey's insights to support particular arguments. A detailed engagement with the entirety of his oeuvre, however, lies beyond the scope of this discussion, though key aspects relevant to the incarnation have been integrated into the main text.

blood."[13] Though brief, this statement encapsulates his conviction that the incarnation cannot be restricted to humanity—any more than it can be confined to males alone. Linzey develops this idea further in subsequent writings, arguing that the Logos became a mammal and that the flesh assumed in the incarnation is not exclusively human but is the same organic flesh and blood shared with other creatures. By invoking the category of "flesh," Linzey deliberately expands the theological scope of the incarnation to embrace all sentient, embodied life, thereby situating non-human animals within the incarnational mystery itself.

While it is indeed the case that the incarnation takes place in human flesh, and that Scripture attributes to humanity a distinctive vocation within creation (Rom. 8:19–23), this need not entail that other creatures participate in redemption merely *through* humanity, nor that their significance is secondary. I agree with Linzey that the Word's assumption of *flesh* (sarx) in John 1:14 signals a solidarity that reaches beyond the human to embrace all embodied life. The incarnation is particular in form but universal in scope: through the assumption of human flesh, the Logos identifies with the entire order of creaturely existence.

This does not dismiss the biblical theme of human responsibility; rather, it reframes it. Humanity's unique capacity for conscious response to the Creator does not imply that animals are passive recipients of redemption mediated solely by humans. Instead, animals are fellow participants in the Spirit-filled economy of redemption, with their own creaturely forms of praise and witness. Humanity's role, then, is not one of domination or exclusive mediation but of *solidary representation*, in which humans join their voices with those of other creatures rather than speaking *on their behalf*.[14] In this sense, animal redemption does not simply "coincide" with human redemption, nor is it entirely dependent on

[13] Andrew Linzey and Dorothy Yamamoto, eds., *Animals on the Agenda: Questions about Animals for Theology and Ethics* (Champaign, IL: University of Illinois Press, 2010), xvi.

[14] I am cautious about framing this role in priestly terms. Linzey's theological vision resists anthropocentric hierarchies that make animals dependent on human mediation. I am grateful to Frank Macchia for his probing questions on this aspect of my argument. His feedback prompted me to nuance more carefully the relationship between the particularity of the incarnation in human flesh and its universal scope for all creation, as well as to articulate more clearly the dynamics between human vocation and creaturely participation in redemption.

it; rather, both are caught up together in the one redemptive movement inaugurated in Christ and effected by the Spirit.

In recent years, biblical scholar Richard Bauckham has drawn attention to the ecological depth of the declaration that "the Word became flesh." While the immediate focus of John 1:14 is on the Word's assumption of human existence—with all its vulnerability, frailty, and mortality—Bauckham urges readers to recognise the broader significance of the term *flesh*. He argues that the phrase "The Word 'became flesh'" encompasses the shared mortal nature of all living beings in order to bestow the eternal life of God upon creation as a whole.[15] Clough turns to John 1:14 to deepen theological reflection on the incarnation. He interprets the verse

[15] Richard Bauckham, *Bible and Ecology: Rediscovering the Community of Creation* (London: Darton, Longman and Todd, 2010), 164. Some biblical scholars have previously asserted that John's Gospel diminishes the value of creation through various dualities. Yet, if we read the prologue as extending beyond humanity, seeing 'flesh' as both human and nonhuman creation, the Gospel becomes a narrative that stretches eternity across a vast spectrum of mortal existence. In this view, human and nonhuman creatures move together toward a shared destiny in Christ. Though Origen is often linked to stoic dualism and known for his indifference to animals, referring to them as 'dumb beasts,' his teachings do not fully discount the value of nonhuman life. He acknowledges that all animals are Spirit-breathed and possess a soul, though he confines the Spirit's work in them to the animating force of life and movement. This, of course, contrasts sharply with his focus on the richer human experience of the Spirit, where the rational soul, with its freedom, can stir transformation through God's Spirit. Origen, *On First Principles* (Notre Dame, IN: Ave Maria Press, 2013), 154. Gregory of Nazianzus rejected Origen's dualistic perspective and instead emphasised a Pneumatology that viewed the Holy Spirit as the only subject capable of free movement and activity. Bergmann, *Creation Set Free*, 261. Bergmann notes that the emphasis on the Spirit's permeation, indwelling and consummation of creation in Gregory of Nazianzus' cosmic trinitarian Pneumatology lends to the idea of a liberated creation. Gregory believed that the Spirit dwells among all creatures, is active in the angels, watches over all things, and brings about movement in living beings. For Gregory the Spirit adds life to everything, and it is the Spirit that bestows life and all that lives so that all life might sing the praises of its Creator. He also recognised that the Spirit is the agent of all life, adding life to everything and bestowing life in a vivifying sense. Nazianzus, *St Gregory of Nazianzus: On God and Christ: The Five Theological Orations and Two Letters to Cledonius*, 55–56. Gregory of Nazianzus invites his audience to observe the behaviour and characteristics of different creatures, such as fish, birds, grasshoppers, swans, and peacocks, and to contemplate the intelligence and beauty displayed in their actions. Gregory prompts his listeners to question the source of their melody, their love of beauty, and their skills in work and art. He suggests that these qualities are not arbitrary but are evidence of a Creator who endowed these creatures with unique features and abilities. See Bergmann, *Creation Set Free*, 166. Although Origen and Gregory of Nazianzus have differing perspectives on the role of the Spirit in non-human creatures, both recognise the importance of the Spirit in the life force and movement of animals.

as signifying that the Word embraced "the life-substance shared among humans and other animals," emphasising that the incarnation involves becoming a *creature*.[16] For Clough, the decisive theological distinction is not between humans and other animals but between God and all creatures. Drawing on Barth's Christocentric reading of the *imago Dei*, he identifies Jesus—not humanity in general—as the true image of God. From this perspective, the key distinction lies between Christ and all other creatures, rather than among creatures themselves. Clarifying this theological structure, Clough argues, provides the necessary foundation for discerning the distinctive ways particular creatures reflect the image of God.

While Clough's analysis aligns with other theologians, particularly in emphasising a cosmic Christology and challenging anthropocentric interpretations of Christology, he distinguishes himself by focusing more directly on non-human animals. Clough observes that even ecotheologians have often overlooked animals, focusing instead on ecosystems and the broader cosmos. Clough adopts a nuanced approach to the relationship between the particularity of Jesus' incarnation and the broader scope of cosmic Christology. On the one hand, Clough explores the significance of a cosmic Christology—one that expands the scope of Christ's work beyond Jesus' specific Jewish, male, and human identity to encompass all of creation. This cosmic view highlights the universal implications of the incarnation and redemption.[17] However, Clough is also critical of any tendency to dismiss the particularities of Jesus' humanity. He expresses scepticism about whether the emphasis on the distinction between God and creatures, or a purely Christocentric interpretation of the *imago Dei*, renders these particularities irrelevant. Instead, Clough holds that both the specificity of Jesus' incarnation and the expansiveness of cosmic Christology are integral to understanding the full scope of God's redemptive

[16] Clough, *On Animals, Systematic Theology*, 84–85. Stephen H. Webb contributes to this perspective, stating that "animals are a part of the incarnation" because traditional Christology combines the divine and the human without excluding any elements from the physical or material aspects of creation. Webb emphasises that humans, as mammals, are flesh and blood creatures with bodily needs and functions. These early expressions by theologians engaged in nonhuman animal issues illustrate the recognition of the potential richness of a "deep Incarnation" theology. See Stephen H. Webb, *On God and Dogs: A Christian Theology of Compassion for Animals* (New York: Oxford University Press, 2002), 170.

[17] Clough, *On Animals, Systematic Theology*, 103.

work. By preserving the tension between the particular and the universal, Clough offers a framework that honours the details of the incarnation without limiting its cosmic significance.

Denis Edwards proposes that God has inscribed an imprint and reflection of divine Wisdom within every creature and throughout creation itself. This means that creatures as diverse as whales, koalas, and humans, each in their distinctive and interconnected ways, serve as reflections of divine wisdom. Edwards highlights the unique human capacity to participate consciously in divine wisdom through our relationships with the natural world, our pursuit of justice and truth, and our contemplation of the Word. He suggests that humanity's responsibility for the well-being of the global community arises not only through the life-giving work of the Spirit but also through participation in this holy Wisdom.[18]

Drawing on Athanasius, Edwards observes that, since the incarnation, God's self-revelation is no longer mediated solely through the imprint of Wisdom in creation but through Wisdom made flesh in Christ. This does not negate the Wisdom embedded in the world; rather, it intensifies it. In light of the incarnation, Edwards argues, we have even greater reason to discern and celebrate divine Wisdom in all its manifestations—whether in ancient trees, delicate wildflowers, endangered species, or human communities.[19] He notes that Athanasius does not sharply separate humanity from the rest of creation but envisions a comprehensive redemptive scope: through the Word made flesh, humans are forgiven, deified, and adopted as beloved children of God, while the rest of creation undergoes its own distinct transformation in Christ.[20]

Edwards' account is especially significant for animal theology because it resists rigid anthropocentric boundaries. Rather than positioning humans as the exclusive bearers of divine reflection, he frames Wisdom as enfleshed, dynamic, and cosmic in scope. His vision of redemption unfolds across three interconnected dimensions. First, at the level of matter, the incarnation initiates the universe's transfiguration, anticipating its glorification and fulfilment. Second, at the level of biological life, the biblical promise of liberation and fulfilment extends to all species and

[18] Edwards, *Deep Incarnation*, 127–128.

[19] Edwards, *Deep Incarnation*, 127–128.

[20] Edwards, 127–128.

individual creatures. Third, for humanity, redemption involves forgiveness, the indwelling of the Spirit, resurrection life, and communion with the whole of creation within the triune life of God.[21]

This theological vision situates non-human creatures not on the margins but within the very fabric of God's redemptive drama. Through the incarnation, God draws all creatures into deeper communion, honouring their unique identities and evolutionary histories.[22] It provides a constructive framework for integrating animals into Christological and pneumatological reflection—one that neither collapses creaturely distinctions nor sidelines their participation in divine Wisdom.

Wallace, similarly, argues that the Christian tradition celebrates the embodiment of God in creatures other than human. He posits that Christianity embodies a double incarnation, whereby God becomes flesh in both human and non-human creatures. Wallace writes, "Just as God became human in Jesus, thereby signalling that human beings are the enfleshment of God's presence, so also becoming avian in the Spirit, God signals that other-than-human beings are also the realisation of God's presence."[23] Wallace expands his understanding of incarnation to include non-human creatures, specifically an avian being that is revealed as hovering over the waters in the creation story and the dove at Christ's baptism. At the core of Wallace's ecopneumatology is a dynamic understanding of the Spirit that extends beyond a passive or static presence. Instead, the Spirit functions as a potent ecological symbol, embodying a semiotic value within creation. Wallace challenges traditional depictions of the Holy Spirit as merely an ethereal "Holy Ghost" by presenting the Spirit as an avian, fully embodied force intertwined with the elemental forms of wind, water, fire, and earth—the foundational components of life itself.[24] This vision portrays the Spirit as actively brooding over creation, not only nurturing but also judging, creating, and even destroying as part of its transformative role. For Wallace, the Spirit is depicted in the Gospels as a divine agent of political and cultural subversion, disrupting

[21] Edwards, 127–128.

[22] Edwards, 127–128.

[23] Wallace, *When God Was a Bird*, 85. See also Swoboda's considerations of Wallace's ecopneumatology: Swoboda, *Tongues and Trees*, 130–41.

[24] Mark I. Wallace, *Finding God in the Singing River: Christianity, Spirit, Nature* (MI: Fortress Press, 2005), 7–8.

established power structures and promoting an egalitarian ethos.[25] This subversive force seeks to deliver life and justice to all creatures, large and small, pointing towards a vision of society grounded in the Spirit's life-giving and unifying power. Wallace argues that Christian animism is not a contradiction but instead reflects Christianity's natural habitat as a spirituality grounded in the ancient belief of an animal God that represents the presence of Spirit in everything, both human and non-human.[26] While Wallace's vision powerfully draws attention to the Spirit's embodied presence in non-human creatures, his notion of a "double incarnation" risks obscuring the singularity and sufficiency of the Logos's incarnation in Jesus Christ. Rather than a second incarnation, I understand the Spirit's presence in creation as ***derivative and participatory***, flowing from and oriented towards the incarnate Christ, who remains the unique mediator of divine life to all flesh.

This section has surveyed key developments in animal theology, focusing particularly on Christological themes of soteriology and incarnation. By attending to the Spirit's work with "all flesh," it has traced how animals are drawn into the divine drama of redemption. The incarnation's embrace of creaturely flesh establishes a theological foundation for discerning the Spirit's activity within and beyond human experience, fostering transformative communion between God and all living beings. Extending the concept of deep incarnation, the discussion now turns to how this theological horizon opens onto the Spirit's redemptive work. The interplay between incarnation and Spirit reveals a more expansive vision—one that encompasses the whole community of creation within God's saving purposes.

[25] Wallace, *Fragments of the Spirit*, 123–25.

[26] Wallace, *When God Was a Bird*, 3. Mark Wallace sheds light on a forgotten thread in Western theology, one that sees the Spirit as a life force, intimately woven into the fabric of nature. Although in theory, the Spirit has always been described as both the Spirit of God and the Spirit of creation, with the former embodying the power of reciprocity within the Trinity and the inner force of human redemption, and the latter as the breath of God that animates and sustains the universe, in practice the emphasis has almost exclusively been on the Spirit of God. As a result, the cosmic role of the Spirit as the life-giving force within all of creation, both human and non-human, has been consistently overlooked. Wallace, *Fragments of the Spirit*, 136.

An Incarnational Pneumatology

The embrace of creaturely flesh in the incarnation lays the groundwork for understanding how the Spirit operates within and beyond human experience.[27] This points towards a transformative communion between God and all living creatures which amplifies the unity of the Spirit and Christ in the ongoing process of incarnation and redemption.[28] In this section, I consider the incarnation as a transformative process that not only shapes human nature but also initiates a comprehensive harmony with the entirety of creaturely existence.[29] The Spirit's involvement, complexly linked with the embodied mediation of Christ, leads to a profound realignment that extends beyond human confines to encompass all living beings. In this context, the connection between the incarnation and the outpouring of the Spirit emerges and positions the incarnation as a profound dimension of Christ's role as the Spirit Baptiser for the entire creation.

A more integrated view of the Spirit's redemptive work can shed further light on the relationship between incarnation and creation. Put simply, the Spirit is not understood solely through the future-oriented promise of redemption but as actively present in creation's ongoing processes, including its suffering. The Spirit is depicted as standing in solidarity with creation, particularly amidst its groaning and travail, as expressed in Romans 8. Such an understanding reframes traditional

[27] Swoboda identifies a weakness in the Pentecostal soteriological framework, noting that it often lacks a coherent theological element that effectively integrates the work of Christ with the Spirit's outpouring at Pentecost. He contends that any understanding of the Spirit's embodiment in creation must be grounded in the creative love of the Word (Christ). Swoboda, *Tongues and Trees*, 105.

[28] While deep incarnation provides the Christological foundation for understanding Christ's identification with all creation, Van Ruler's distinction between Christology and pneumatology highlights the Spirit's unique role in inhabiting and transforming creation. As Van Ruler argues, inhabitation is pneumatology's defining category, ensuring that the Spirit's work complements and extends the redemptive implications of incarnation. This distinction affirms the dynamic and relational nature of the Spirit's engagement with all flesh, ensuring that the theological scope of salvation fully encompasses non-human creatures. Arnold A. van Ruler, *Calvinist Trinitarianism and Theocentric Politics: Essays Toward a Public Theology*, trans. John Bolt, vol. 38, Toronto Studies in Theology (Lewiston, NY: The Edwin Mellen Press, 1989), 27–46.

[29] Macchia, *Jesus the Spirit Baptizer*, 126–27.

soteriological perspectives, offering a broader vision of salvation that encompasses all of creation, including non-human creatures.

Frank Macchia offers a distinctive and generative perspective on the interrelation of the Word, the Spirit, and flesh within the mystery of the incarnation. Central to his account is the notion of "fleshly witness" as a sacrament of the Spirit, grounded in the profound unity between the divine Son and his humanity. This unity, he argues, becomes the conduit through which the Spirit is imparted to all flesh. Macchia frames the incarnation not merely as a redemptive act directed towards humanity but as a transformative process that sets in motion a comprehensive harmony encompassing the whole of creaturely existence. The Spirit, intimately bound to Christ's embodied mediation, catalyses a deep realignment that reaches beyond human boundaries, drawing all living beings into its embrace. This nexus between the incarnation and the Spirit's outpouring forms a pivotal strand in Macchia's theology, positioning the incarnation as a vital dimension of Christ's role as Spirit Baptiser for the entirety of creation.

For Macchia, the Word imparts the Spirit to flesh through a fleshly medium—Christ's humanity itself—which functions as the self-declaration of the Son and, in turn, as a sacrament of the Spirit. This uniquely human medium participates in the transmission of the Spirit from the Father to all flesh, serving as the representative locus through which the human story is rewritten. While the incarnation remains a singular event—the Logos is uniquely present in Jesus Christ—the diverse expressions of all flesh are embraced through the universal and cosmic indwelling of the Spirit, active since the beginning (Gen 1:2).[30] Through the Spirit's impartation, the incarnate Son incorporates all flesh into himself, shaping it in his image. Through the Spirit's impartation,

[30] Macchia, *Jesus the Spirit Baptizer*, 126–127. While Sánchez has humans primarily in view, he explains that while the Son's primary task is the incarnation (hypostatic union), the Spirit's specific role is to sanctify human nature through this indwelling. The Holy Spirit's distinctive function extends to the consecration of the human nature of the Son and subsequently to the holiness of God's adopted sons, who receive the same Spirit through grace mediated by the Son. Leopoldo A. Sánchez, *Receiver, Bearer, and Giver of God's Spirit: Jesus' Life in the Spirit as a Lens for Theology and Life* (Eugene, OR: Wipf and Stock Publishers, 2015), 96. This understanding of the Spirit's role in sanctifying the Son's humanity provides a framework for considering how Christ, as a sacrament of the Spirit, extends this sanctifying work beyond humanity to all flesh. By mediating the Spirit through his creaturely flesh, Christ also draws non-human creatures to himself, thus affirming their place within the scope of salvation.

Macchia argues, the incarnate Son incorporates all flesh into himself, shaping it in his image. The fleshly medium, participating in divine self-impartation through the Son and in the Spirit, must be so closely united with the divine Son that it fully engages in this divine incorporation. Essential to this position is the unity of identity (or person) between the Son and his humanity—essentially, the concept of incarnation. Macchia posits that such a fleshly witness, intimately linked to the divine Son's Spirit impartation, becomes a conduit, turning the embodied existence of Jesus Christ into a sacrament of the Spirit for all creation.[31]

For Macchia, the renewal of creation through the divine Spirit is intertwined with an embodied mediation that is particularly emphasised in the gospel. Here, the divine Son assumes human form, providing an avenue for humanity to partake in the divine life, and this incarnational process finds its culmination at Pentecost.[32] Macchia posits a crucial link between the incarnation and Pentecost, affirming that this connection is achieved through the mediation of a life baptised. From the depths of Jesus's being, living water pours forth, baptising all flesh in the Spirit (John 1:33; 7:38). This link between incarnation and the outpouring of the Spirit places Christ's role as the Spirit Baptiser within the sweeping scope of creation's renewal.[33] The significance of the Jordan River in this context lies in the implication that the incarnation is a complex event involving both the

[31] Macchia, *Jesus the Spirit Baptizer*, 126–127.

[32] Macchia, *Jesus the Spirit Baptizer*, 124. Linzey has also noted the connection with the baptising Spirit on all flesh and the outpouring of the Spirit at Pentecost, see Andrew Linzey, *Christianity and the Rights of Animals* (Eugene, OR: Wipf and Stock, 2016), 78–79.

[33] Lamp, who also discusses this idea of a Spirit-baptised creation, captures how Pentecost provides a unique perspective on the Spirit's presence in the new creation. Lamp would agree with my thinking here, as he also notes how Frank Macchia further expands this perspective by reimagining the Pentecostal experience of Spirit baptism, moving beyond a human-centric framework to encompass the entirety of creation. Lamp, Jeffrey S. "Incarnate Pneumatology: Thinking Ecologically about the Spirit." Presented at the 45th Annual Meeting of the Society for Pentecostal Studies, 2016, 11.

Drawing inspiration from Gregory of Nyssa, Macchia envisions the Spirit as a living and substantial kingdom, with Christ as its anointed and ruling monarch. According to Macchia, this perspective implies that Christ, as the King, reigns over all creation through the Spirit, who permeates every aspect of the created order. Spirit baptism, as Macchia posits, involves the imparting of the Spirit into the world by the Son, thereby liberating creation from historical constraints and opening up new possibilities for an eschatological existence. From an ecotheological perspective, Lamp proposes that the signs of the Spirit's outpouring, as described in both Joel and Acts 2, include not only manifestations

Word and the Spirit. The Word assumes flesh through the agency of the Holy Spirit, with the Spirit's mediation at the incarnation performing a sanctifying role, as seen in the conception of Jesus as the holy Son of God. Moreover, Macchia recognises Christ as the mediator of the Spirit's work, particularly in the role of a Spirit Baptiser. This mediation extends beyond the incarnation, as the outpouring of the Spirit is anticipated by the incarnate Christ. While the incarnation is a singular and unique event, deeply rooted in the whole of evolutionary history, the outpouring of the Spirit is cosmic in scope. I find Macchia's emphasis on the unity of person between the Son and his humanity a compelling way of understanding how the embodied life of Jesus becomes the sacramental site through which the Spirit's life is communicated to creation. Yet, as my broader argument suggests, this process must also be read through the lens of

among humans such as dreams and visions but also natural phenomena—both cosmic and terrestrial.

The concept of creation as an active and significant character in Joel 1–2, receiving the word of the Lord and lamenting for salvation, aligns with the insights of Griffiths', who argues for a broad ecological understanding of the term "*basar*." Griffiths notes that the promise of God's Spirit in Joel 2:28 is poured out in terminology as wide as creation, which suggests that Joel may have intended to include all of creation in this outpouring. This interpretation is supported by the LXX's translation of Joel 2:28, where "σάρξ" is understood to encompass non-human flesh, as noted by Griffiths. Griffiths further argues that this ecological understanding of "σάρξ" in Acts 2:17 implies a broader relationship beyond humanity. Thus, Griffiths' insights suggest that both Joel and Luke likely understood the outpouring of the Spirit of God as extending to all of creation, including non-human flesh. See John D Griffiths, "Spirit-Baptised Creation: Locating Pentecost in the Meta-Narrative of Creation and Its Implications for a Pentecostal Ecology," *Australasian Pentecostal Studies* 22, no. 1 (June 9, 2021): 55. Griffiths' insights suggest that both Joel and Luke likely understood the outpouring of the Spirit of God as extending to all of creation, including non-human flesh. This expansive understanding of the Spirit's activity naturally connects to the cosmic imagery described in Acts, where the sun turning to darkness, the moon to blood, fire, and smoky mist signify a universal impact. Lamp interprets these phenomena as metaphors that evoke the Spirit's creative work in Genesis, where the divine wind hovered over primordial chaos, suggesting a continuity between creation and Spirit-filled renewal. Lamp, "Incarnate Pneumatology," 10–11. In Acts 2, the Spirit is portrayed as a wind hovering over a broken creation, anticipating a new creation. For Lamp, just as the Spirit was present to bring life to the original creation, the Spirit now brings life to the new creation. In his expansion of this idea, Lamp notes how Pinnock emphasises the "cosmic operation of the Spirit" in both creation and redemption, portraying the Spirit as the ecstasy of divine life that births the universe and strives towards complete unity. Similarly, Leonardo Boff describes the Holy Spirit's work as "pneumatising" creation, redeeming it from death and reintegrating it into the life of the Trinity. See Lamp, "Incarnate Pneumatology," 12.

the Spirit's dynamic work beyond the human sphere, encompassing non-human creatures within a Spirit-baptised economy of praise and renewal. Macchia's theology provides a fertile starting point for envisioning this cosmic scope, though it requires further pneumatological development to fully address its implications for animals and the wider creation. While he does not explicitly engage with non-human creatures, I argue that his framework carries significant, and as yet untapped, potential for advancing animal theology.

The foundational embrace of creaturely flesh in the incarnation sets the stage for comprehending the Spirit's operation within and beyond the human experience, pointing towards a transformative communion between God and all living creatures. This connection underscores the unity of the Spirit and Christ and establishes a narrative of incarnation and redemption that extends to the entire created order. Macchia draws a vital theological thread between the incarnation and Pentecost, the Word and the Spirit. In his account, the Spirit's mediating role within the incarnation carries a sanctifying significance that reaches beyond the human sphere.

Macchia's insights can be deepened by turning to Matthias Wenk's reading of Romans 8, where the Spirit's role in creation's redemption is brought into sharper focus.[34] In the chapter's first 17 verses, the apostle Paul unveils a conception of justification and the formation of a new people by the Spirit, transcending legal constraints to include both Jews and Gentiles. This new community, formed and sustained by the Spirit, becomes the agent and sign of creation's ultimate liberation. The shift in focus from individual believers to the cosmic dimension is noteworthy where the Spirit's renewing and life-giving power extends to the entirety of creation. Paul shifts the horizon from the individual to the cosmic, presenting the Spirit's renewing, life-giving power as extending to the whole of creation. This vision disrupts narrow, individualised soteriologies by situating salvation within a cosmic framework, where the redemption experienced by the children of God prefigures the eschatological renewal of all creation.[35]

[34] Matthias Wenk, "An Incarnational Pneumatology Based on Romans 8:18–30: The Spirit as God's Solidarity with a Suffering Creation," *Religions* 13, no. 3 (2022): 191, https://doi.org/10.3390/rel13030191.

[35] Wenk, "An Incarnational Pneumatology," 3.

Wenk observes that the phrase "first-fruits of the Spirit" can be understood in two ways: referring either to the Spirit itself as the first fruit or to the eschatological fruit that the Spirit brings forth in the lives of believers. In either case, the Spirit is not merely a down payment on future glory but an active agent who begins to cultivate that future in the present—stirring hope, interceding in human weakness, and groaning in solidarity with the whole creation.[36] Paul's vision in Romans 8 refuses to limit the Spirit's work to individual believers. Instead, the Spirit's mission permeates the entire created order, breathing life and hope into a world weighed down by suffering.

As Wenk carefully unpacks Romans 8:18–30, he draws attention to a pneumatological thread running through Paul's soteriology. In this passage, Paul shifts the focus of God's solidarity with a suffering creation from the crucified Jesus to the Spirit.[37] Personal liberation and the redemption of the wider world are depicted as inseparable, each bound together in a shared movement towards ultimate renewal. Wenk draws connections to Old Testament descriptions of eschatological renewal through the outpouring of the Spirit, particularly in Isaiah 32.15–18 and Ezekiel 36.25–35. Significantly, he suggests that limiting the Spirit's role to transformative power overlooks the Spirit's enduring, even grieving, attributes, which point to the Spirit's active presence in confronting and overcoming impediments to life.

Building on this pneumatological horizon, I suggest that the Noahic flood, read through 1 Peter 3:20–21, offers a typological precedent for understanding creation itself as undergoing a purgative and recommissioning passage. The dove as a pneumatophanic sign resonates with Christ's baptism, providing a sacramental pattern that frames creation as Spirit-baptised. This typology underscores the cosmic breadth of the Spirit's activity, consecrating creation itself for renewal and anticipating the eschatological outpouring on "all flesh." Such a reading expands

[36] Wenk, "An Incarnational Pneumatology," 21. Wenk explains that Rom 8:18–30 is centred on the Spirit's solidarity and intercession for a suffering creation that longs for liberation. Paul here is not only concerned with personal character transformation, but also with the eschatological liberation of creation, leading to newness of life in fellowship with God's children (Rom8:21). Although the text does not provide explicit ethical instructions, the broader Pauline context suggests that believers are encouraged to participate in the Spirit's mission in the world, as "the church is to be where the Spirit is, in solidarity with the suffering world." Wenk, "An Incarnational Pneumatology," 22.

[37] Wenk, "An Incarnational Pneumatology Based on Romans 8.18–30," 5.

the scope of pneumatology by framing the entire created order as Spirit-baptised. This sacramental typology anticipates the eschatological outpouring on "all flesh," situating non-human creatures not at the margins of redemptive history but within its very flow. It provides theological scaffolding for understanding how the Spirit's renewing work extends to the whole of creation, aligning baptismal imagery with cosmic transformation.[38]

Wenk sees in Romans 8:18–30 a subtle layer often missed in conversations about the Spirit's transformative work—the call to stand with a world in pain, to share in its groaning. The Spirit is not just a promise of salvation but a force in motion, actively guiding all of creation, human and non-human alike, towards its redemption.[39] The Spirit of Christ actively contributes to providing new life to the entire creation, liberating it from potential extinction.[40] Wenk pushes for a wider lens on soteriology, imagining a "soteriology of the Spirit" that reaches beyond the legalistic frameworks often found in Western theology. He argues for the inclusion of resurrection, renewal of life—particularly in its ecological sense—and reconciliation as key threads for a more complete view of salvation. Such inclusion, he argues, broadens the church's soteriological vision, aligning more accurately with God's passion for the world.[41]

Together, the perspectives of Macchia and Wenk allow for a richer understanding of incarnational pneumatology. They open the way for a broader soteriology that includes all creation, thus offering significant implications for developing a theology that fully considers the role of animals within God's redemptive plan.

[38] I am grateful to Amos Yong for his probing questions and insightful engagement on this aspect of my argument, which prompted me to clarify and develop the pneumatological implications of this typology more fully.

[39] Wenk, 5.

[40] Edwards would concur with Wenk, stating that in the word made flesh, God embraces the whole labour of life on the earth, with all its evolutionary processes, including death, predation, and extinction, in an event that is both a radical identification of in love and an unbreakable promise. Edwards, "Redemption of Animals in an Incarnational Theology," 95.

[41] Wenk, "An Incarnational Pneumatology Based on Romans 8.18–30," 5.

Animals as All Flesh Partners

This section extends the preceding theological reflections on the Spirit's role in creation and redemption by turning to the place of animals as active participants in God's purposes. Rather than casting animals as passive elements within the created order, I emphasise their inclusion as "all flesh" partners in the Spirit's life-giving work. Recognising animals as recipients of divine grace and participants in the Spirit's ongoing activity reframes their theological status: their redemption is significant in itself, yet it is accompanied by their active participation in the Spirit's work of eschatological renewal. In this way, animals embody and reflect the Spirit's sustaining presence, bearing witness to a divine economy that encompasses the entirety of creaturely life.

A. J. Swoboda creatively coins the term *Eco-Glossolalia* to symbolically refer to a Spirit-baptised creation, or a baptism of "all flesh."[42] Swoboda's approach broadly unites humanity with creation as eschatological partners through the empowerment of the Spirit and the outpouring of charismatic gifts. The distinctive Pneumatological strengths of Pentecostal theology permit the embedding of ecotheology into a Pentecostal worldview. Within the context of his Spirit-baptised creation metaphor, Swoboda renders Pentecostal Pneumatology and its definition of Spirit baptism simply as "immersion."[43] This immersion includes powerful experiences of encounter with the living God. It includes healing, empowerment for witness and eschatological preparation, all of which have implications for the community of faith. Swoboda argues that the holistic Spirit helps "conceptualise both the role of the individual in the church and the role of every partner in creation" to see "the church and creation as holistic bodies that are interdependent and interdynamic."[44]

[42] A. J. Swoboda, "Eco-Glossolalia: Emerging Twenty-First Century Pentecostal and Charismatic Ecotheology." *Rural Theology* 9, no. 2 (2011): 103; Swoboda's metaphor is constructed on the communal understanding of Pentecost (as described in Acts 2:17), where the Spirit baptises "all flesh." Swoboda clarifies that his thesis of a Spirit-baptized creation is developed from the creational narratives of Genesis 1–3, a portrait "of God's Spirit giving life and taking life in every element of the created realm." A J Swoboda, *Tongues and Trees: Towards a Pentecostal Ecological Theology* (Dorset: Deo Publishing, 2013), 241.

[43] Swoboda. *Tongues*, 241.

[44] Swoboda. *Tongues*, 221.

Swoboda's approach is fertile but not without challenges. He acknowledges the tension between his metaphor and the New Testament witness, where Spirit baptism is narrated primarily with ecclesial (rather than creational) purposes in view. This tension is precisely where a Pentecostal pneumatology has the capacity to expand.[45] If, as Frank Macchia argues in his "Creation Pneumatology," the entire earth longs and groans for the liberty of the Spirit, then the Spirit's transformative work is not confined to the church but extends towards the comprehensive renewal of creation itself.[46] Macchia describes this as creation's transformation into "the dwelling place of the Spirit in the image of the risen Christ."[47] In other words, the charismatic and the cosmic are not competing horizons but deeply intertwined dimensions of the Spirit's mission.

Amos Yong develops this vision with particular attention to non-human creatures, pressing the biblical language of the Spirit poured out on "all flesh" into genuinely expansive territory. He uses the example of the pneumatological vision of the Prophet Isaiah and links "the charismatically anointed Messiah with the healing and reconciliation of creation's

[45] Swoboda, *Tongues*, 241. While Swoboda explains that Spirit baptism is unique to Christian experience, he does not entirely disqualify a shared experience by non-human agents. Swoboda suggests that the life-giving power of God's Spirit extends to both humans and non-human creatures, highlighting the need to tread carefully when considering the concept of Spirit baptism ecotheologically. He notes that Spirit baptism has often emphasised a unique separation of humans from the rest of creation, positioning baptised believers as distinct. However, Swoboda also sees a transformative potential in Spirit baptism—it could open believers to a new understanding where the Spirit-baptised church participates in a renewed creation. In this sense, Spirit baptism has a dual potential: while it may initially emphasise a separation between the baptised and the natural world, it also holds the promise of bringing believers into a new paradigm where they engage in God's broader eschatological renewal of all creation. Swoboda, *Tongues*, 203.

[46] Macchia, "The Spirit of Life," 129.

[47] Macchia, "The Spirit of Life," 129.

destroying forces."[48] This includes the Spirit poured out on all creaturely flesh, including the wolf and the lamb, the leopard and the kid, "all of whom are included in the blessings of God promised under the covenant of Noah."[49] Here, the Noahic covenant becomes the theological hinge: God's promise encompasses *every living creature*, as David Clough also underscores, ensuring that animals are not merely bystanders to divine address but recipients of it, stating, "humans are not the only creatures that are addressed by God and called to live lives in response to God."[50] In this light, the Spirit's activity cannot be confined to human communities; rather, it extends in groaning solidarity with all creation, interceding on its behalf (Rom 8) and standing as the divine pledge that the future of creation (though vulnerable to decay) will not be forsaken.[51]Yong's pneumatological horizon powerfully widens the field of theological imagination, compelling us to reckon with animals not as peripheral beneficiaries but as participants in the Spirit's redemptive mission.

Yong's expansive pneumatological vision opens up a horizon in which the charismatic and the cosmic are not competing but convergent realities. By emphasising the Spirit's outpouring on "all flesh," Yong gestures towards a theology that holds together creaturely diversity, charismatic empowerment, and eschatological hope. Building on this, Moltmann

[48] Yong, *The Spirit Poured Out on All Flesh,* 300. The precise wording used in Isa 66:23 and Ps 145:21, which includes the Hebrew phrase *kol basar* (meaning "all flesh), can be understood to refer to all living creatures, including humans, animals, or all of mankind. It is unlikely that these verses exclusively refer to animals or to the physical human body alone. This leaves two possible interpretations: either all humans or all humans and animals. In the Hebrew Bible, the use of *kol basar* to refer to both humans and animals is not uncommon and is found in various passages (such as Gen 6:13, 17; 9:11, 15, 16, 17; Leviticus 17:14; Numbers 18:15; Psalm 136:25). In fact, some argue that when animals and humans are intended to be referred to together, the phrase "all flesh" is typically used. J. H. Eaton, *Psalms: Introduction and Commentary* (London: SCM Press, 1967), p. 316.

[49] Yong, *The Spirit Poured Out on All Flesh*, 300.

[50] Clough, David L. *On Animals, Systematic Theology*, 41.

[51] Lamp connects the Spirit's outpouring on "all flesh" in Joel's language, raising the question of whether more than just human beings are included in this divine outpouring. Lamp argues that the creational language in the passage, particularly the mention of "all flesh," may imply a broader scope, especially considering Paul's usage in 1 Corinthians 15:38–39, which extends the concept to include non-human creation. Lamp, Jeffrey S. "Incarnate Pneumatology," 10.

develops the cosmic implications of such a vision with striking eschatological clarity. Whereas Yong foregrounds the Spirit's dynamic presence within creation's ongoing life and groaning, Moltmann casts his gaze towards the ultimate destiny of all creatures. His theology stretches the Spirit's activity across the entire fabric of creation, envisioning an eschatological communion in which every living being participates in the Spirit-led renewal of the world.

Moltmann's exposition of "all flesh" extends beyond the confines of human boundaries, encompassing all living beings without discrimination.[52] This notion finds resonance in his portrayal of the ultimate manifestation of God's Spirit, whereby humanity undergoes a transformative metamorphosis, collectively becoming a prophetic community. The dynamic effect of the outpouring of God's Spirit reverberates throughout the entire fabric of existence, catalysing not only the renewal of diverse life forms but also the revitalisation of the interwoven community of living creatures that inhabit the Earth. Moltmann also envisions all creatures as part of the "all flesh" that will share in resurrection.[53] Ryan Patrick McLaughlin draws out the implications of this vision for animals, highlighting Moltmann's concept of a cosmic community that must be upheld by laws safeguarding the dignity and integrity of every creature. If animals are part of this "all flesh," McLaughlin argues, then they stand before God as creatures of inherent worth, fully entitled to share in the transfiguration of creation.[54]

Moltmann suggests that the encounter with God, ignited by the Spirit, stretches far beyond the boundaries of individual humans or species. His view embraces the fullness of creation and sees "all flesh" being brought into communion with God.[55] This expansive vision reflects Moltmann's

[52] Moltmann, *The Spirit of Life*, 57.

[53] McLaughlin, "Anticipating a Maximally Inclusive Eschaton," 26. This will be addressed in more depth in Chapter 6.

[54] McLaughlin, "Anticipating," 30.

[55] Moltmann, *The Spirit of Life*, 57. Moltmann, reflecting on the Noahic covenant, interprets the phrase "all flesh" as an inclusive term that extends beyond humanity to encompass all living beings. He further argues that divine justice, which the Messiah is to establish, will also restore and enliven the land itself, drawing on Isa 32:15-18: "When the Spirit is poured upon us from on high, the wilderness will become a fruitful field, and the fruitful field will be deemed a forest. Then justice will dwell in the wilderness, and righteousness abide in the fruitful field. And the effect of righteousness will be peace…" For Moltmann, this scripture resonates as a continuation of the Pentecost experience,

understanding of the Spirit's work, which reaches into every form of life, echoing in the deepest parts of creation. He situates this vision within the biblical narrative itself. The covenant with Noah becomes a key theological anchor—a sign of God's universal promise to all living creatures. This covenantal framework aligns with the Psalms' depictions of divine providence sweeping across the whole of creation, sustaining and caring for every creature. For Moltmann, the encounter with God through the Spirit is not limited to human hearts or human history. It pulses through the entire created order, echoing in the life of every creature and drawing all flesh into communion with God.[56]

Moltmann's eschatological vision thus casts the Spirit's work across the entirety of creation, gathering all living beings into a future communion shaped by divine promise. His grounding of this vision in the Noahic covenant provides a theological anchor, reminding us that God's concern has always encompassed "all flesh." This covenantal horizon,

where the Spirit of life is poured out upon "all flesh"—meaning not only human life but all living creatures. Jürgen Moltmann, "The Liberation of the Future and Its Anticipations in History," in *God Will Be In All: The Eschatology of Jürgen Moltmann*, ed. Richard Bauckham (MN: Fortress Press, 2001), 284. Cunningham notes the biblical recognition of different kinds of flesh, such as that of cold-blooded and warm-blooded animals. While all creatures share in having fleshly bodies, distinctions exist within the category of flesh. This mirrors empirical observations, allowing for the identification of shared aspects with other species while acknowledging differences. For Cunningham, the fall affects all creatures, placing "all flesh" under divine judgment. He notes the cosmological significance of the fall, indicating a broader disturbance in the created order, underscoring the need for healing in "all flesh." See David S Cunningham, "The Way of All Flesh: Rethinking the Imago Dei," in *Creaturely Theology: On God, Humans and Other Animals*, ed. Celia Deane-Drummond and David Clough (London: SCM Press, 2009), 114.

[56] David Cunningham also notes that during the decision to flood the earth, the focus extends beyond humanity to encompass all flesh. The covenant post-flood is not exclusively with humans but includes all flesh, emphasising a broader divine connection with all creaturely life. The term "flesh" facilitates a nuanced understanding of both commonality and differentiation among living beings. He identifies the biblical recognition of different kinds of flesh, such as that of cold-blooded and warm-blooded animals. While all creatures share in having fleshly bodies, distinctions exist within the category of flesh. This mirrors empirical observations, allowing for the identification of shared aspects with other species while acknowledging differences. For Cunningham, the fall affects all creatures, placing "all flesh" under divine judgment. He notes the cosmological significance of the fall, indicating a broader disturbance in the created order, underscoring the need for healing in "all flesh." See David S Cunningham, "The Way of All Flesh: Rethinking the Imago Dei," in *Creaturely Theology: On God, Humans and Other Animals*, ed. Celia Deane-Drummond and David Clough (London: SCM Press, 2009), 114.

however, is not unique to modern theological reflection. Centuries earlier, Ambrose of Milan had already drawn the Noahic covenant into a richly symbolic framework, reading it through the lens of baptism. His interpretation brings sacramental depth to the ecological themes that Moltmann develops, foregrounding the Spirit's transformative power within the natural world.

For Ambrose, the flood signifies more than judgement; it becomes a baptism for the earth itself, a purgative and renewing event that anticipates creation's restoration. The ark and the cross function as parallel instruments of salvation: "Noah was saved by wood, as you also are saved by the wood of the Cross."[57] The dove released by Noah represents the Holy Spirit, prefiguring the Spirit's role in baptism as the agent of renewal and restoration. Building on Ambrose's theological vision, the flood can be understood as a baptism not only for humanity but for the entirety of creation, extending God's covenantal promise to every living creature. In this way, the Spirit's work through the flood narrative points towards a cosmic baptism—a consecration of "all flesh" within the divine economy of renewal.

Conclusion

In this chapter, I have sought to articulate a more inclusive soteriology—one that brings the Christological emphasis on deep incarnation into conversation with the pneumatological vision of the Spirit poured out on all flesh. This integration provides a theological framework that affirms animals as Spirit-baptised creatures who share in the hope of cosmic renewal. By introducing a pneumatological dimension into Christological discourse on animals, this chapter has advanced a more expansive theological vision: one that positions animals not at the margins but as integral participants in the divine drama of redemption.

Engaging the concept of deep incarnation has shown that Christ's identification with creation encompasses not only humanity but the entire order of creaturely existence. Building upon this, an incarnational pneumatology offers a vital complement by emphasising the Spirit's dynamic role in sanctifying, sustaining, and ultimately transforming creation. Together, these Christological and pneumatological perspectives weave a

[57] Ambrose, *On the Mysteries and The Treatise on the Sacraments*, Book 1, Ch 3.

holistic vision of salvation that resists anthropocentric reduction, affirming the dignity and intrinsic worth of non-human creatures. Drawing on Matthias Wenk's interpretation of Romans 8 and Frank Macchia's account of the Spirit as both sanctifier and liberator, the chapter has highlighted the Spirit's solidarity with a groaning creation. The Spirit's redemptive work is not confined to humanity but extends across the entire fabric of life, acknowledging animals as active participants in God's unfolding purposes.

This trajectory naturally leads into the eschatological horizon, where the Spirit's work in creation and redemption reaches its consummation. The outpouring of the Spirit on "all flesh" is both a present reality and a future promise, anticipating the liberation of creation from its bondage to decay. Scriptural imagery—from Isaiah's vision of a peaceable kingdom to the universal restoration in Revelation—underscores that animals are not incidental beneficiaries but co-participants in the hope of new creation. Eschatology, then, becomes the horizon within which the Spirit's redemptive work finds its fulfilment, bridging the present groaning of creation with the promise of its ultimate glorification.

CHAPTER 5

Eschatology

Eschatology, often described as the study of "last things," reaches far beyond the final chapters of human history. It is, at its heart, a vision of renewal—an unveiling of God's ultimate purposes for the whole cosmos, where divine justice and restoration sweep across human and non-human creation alike. Far from being a doctrine concerned only with endings, eschatology opens a window onto the world as it is meant to be: healed, reconciled, and gathered into the fullness of divine life.

This chapter begins by unsettling certain inherited frameworks that have drawn a sharp line between the present world and its eschatological future. Theologies shaped by apocalyptic imagery have often leaned towards narratives of destruction and replacement, eclipsing the value of non-human creatures in the process. Against this, a growing chorus of theological voices offers a different imagination. They envision a future that grows out of the present rather than discarding it, one in which the redemption of creation includes animals as integral participants rather than expendable extras. These perspectives invite us to rethink eschatology not as a distant abstraction, but as a theological vision capable of fostering harmony, justice, and kinship within creation—especially in the face of animal suffering and exploitation.

D. Rizzo, *Animal Pneumatology*, The Palgrave Macmillan Animal Ethics Series, https://doi.org/10.1007/978-3-032-18723-9_5

Orienting Eschatology as Animal-Inclusive

Historically, Christian eschatology has marginalised or excluded animals, often portraying them as symbolic or peripheral to God's redemptive plan. While scriptural visions such as Isaiah's peaceable kingdom (Is. 11:6–9)—where the lion and the lamb coexist peacefully—offer glimpses of an inclusive eschatological future, classical doctrines have overwhelmingly centred on human redemption, sidelining the broader creation. An anthropocentric lens has seen animals as passive elements within creation, disregarding their inherent worth and potential inclusion in God's eschatological purposes. However, I challenge this exclusionary paradigm, drawing from ecotheology, charismatic and Pentecostal traditions, and broader theological movements to affirm the Spirit's transformative work within all creation, including animals.

I argue that eschatology, traditionally understood as the consummating work of the Spirit, provides a paradigm for reimagining the place of animals in God's redemptive plan. Through a pneumatological lens that conceptualises the Spirit as life-giver, advocate, and reconciler, Pentecostal theology offers fertile ground for developing an eschatological vision that includes animals not merely as passive recipients but as active participants in the renewal of creation. This vision deepens when the Spirit's role in consummation is traced alongside the inherent value and agency of all living creatures within God's salvific purposes. The Spirit's redemptive work is not limited to human flourishing but encompasses the liberation, healing, and renewal of all life forms. Isaiah's prophetic imagery of peace among all creatures illustrates this holistic vision of creation's future. This inclusive approach not only critiques traditional anthropocentric eschatologies but also affirms a theological vision in which animals hold value within the eschaton.

Critique of Discontinuity in Eschatology

The critique of discontinuity within eschatological perspectives has gained traction among theologians like Clough and Edwards, who argue that an overemphasis on the separation between the present world and the new creation undermines a holistic vision of redemption. Clough sustains the importance of the Holy Spirit's role in the redemption of not only humanity but also non-human creatures.[1] Affirming the resurrection of non-human creatures safeguards theological reflection on the well-being of all creation within the present order. By considering resurrection as an extension of God's redemptive work, continuity appears between the current creation and the new creation.

This approach aligns with broader challenges to the traditional dualistic framework that privileges the "spirit-man" over "body-nature," which often results in a disregard for the material world. Such dualistic thinking has historically shaped eschatologies that envision the end times as an annihilation of the earth and its creatures, a perspective that feminist theologians, among others, have contested, rejecting any notion of salvation achieved through a form of cosmic destruction.[2] Such a devaluation of the physical is seen as both theologically limiting and environmentally damaging.

Sheila McGinn provides a feminist reinterpretation of Pauline eschatology that shifts the focus from destruction to liberation. Paul's eschatological vision, McGinn argues, does not foretell cosmic devastation but anticipates an unfolding reality in which creation reaches its full potential. From this view, humanity's salvation is intimately tied to creation's fulfilment. This suggests an interdependent eschatology in which both humans and the natural world await liberation as God's children. This interpretation supports a vision of eschatology that is not only inclusive but also protective of the significance of animals within creation.[3]

In contrast, early evangelical and Pentecostal eschatologies, especially those shaped by revivalist and rapture-focused traditions, utilised apocalyptic themes to instil a sense of urgency. These traditions frequently employed imagery of environmental catastrophe to evoke the need for repentance, positioning creation as a space tainted by sin and therefore destined for divine judgement.[4] Such portrayals were often leveraged in homiletic settings. The "old life" of sin was contrasted with the "new

[1] Clough, *On Animals, Systematic Theology*, 169–70.

[2] Sheila E. McGinn, "All Creation Groans in Labour: Paul's Theology of Creation in Romans 8:18–23," in *Earth, Wind and Fire: Biblical and Theological Perspectives on Creation*, ed. Carol J Dempsey and Mary Margaret Pazdan (Collegeville, MN: Liturgical Press, 2004), 122.

[3] McGinn, "All Creation", 122.

[4] Maier, "Green Millennialism," 256. See also Dyer, "When Is the End Not the End?"; and Radford Reuther's chapter on *Religion Narratives of World Destruction*, in Rosemary

life" offered through salvation, reflecting a dispensationalist outlook that framed ecological degradation as either a symptom of human sin or a form of God's impending punishment. Rooted in reformed and dispensationalist theologies, these perspectives frequently discouraged engagement with environmental concerns, viewing them as distractions from the gospel mission. Traditional dispensationalist beliefs perceived ecological interventions as contrary to God's plan for an apocalyptic renewal of creation. As a result, the idea of environmental conservation was often dismissed, with emphasis placed instead on missions and personal salvation.[5] Moreover, a literal reading of biblical texts, such as Peter's vision in Acts, was often invoked to justify the unrestrained use of natural resources, interpreting dominion as a mandate for consumption without limitation. Animals and other elements of nature were perceived primarily as resources for human benefit. This supported an ethos of limitless growth underpinned by the assumption that resources would suffice until the second coming of Christ.[6]

These positions call for a fundamental reimagining of eschatology within contemporary ecotheological discourse. What is needed is not simply a revision, but an alternative vision—one that foregrounds continuity and recognises the Spirit's redemptive activity as encompassing the whole of creation. Such a vision affirms an inclusive eschatology attentive to the flourishing of both human and non-human life. In this way, it cultivates a creation-honouring theological imagination that resists dualistic and apocalyptic extremes, offering instead a vision of eschatological

Radford Ruether, *Gaia and God: An EcoFeminist Theology of Earth Healing* (New York: Harper One, 1992), 61–84.

[5] Maier, "Green Millennialism," 253.

[6] Maier, 253. Similarly, within Pentecostal theology, Shane Clifton critiques the barriers preventing a robust ecological ethic, suggesting that Pentecostalism's theological foundations have the potential to inspire significant ecological engagement. He identifies factors such as premillennial eschatology, "young earth creationism," prosperity theology, and a dualistic, narrow view of the church's mission and argues that Pentecostalism's core theological symbols hold the potential to inspire and sustain a robust ecological practice, provided there is a comprehensive revisioning of the Pentecostal worldview to include an earth-affirming theology of creation. Clifton specifically advocates for a pneumatological theology of creation, proposing a reframing and expansion of the Pentecostal "full gospel" narrative to incorporate an ecological perspective, presenting Jesus not only as a healer of sick bodies but also the healer of a wounded creation. Clifton, "Preaching the 'Full Gospel' in the Context of Global Environmental Crisis," 120–21. For more on Pentecostals and the Creation-Evolution dialogue, see also King, "Evolving Paradigms."

transformation that embraces the entirety of God's creation, animals included.

Shifting Pentecostal Eschatology: From Apocalypticism to Creation Care

A re-orientation within the tradition of Pentecostal eschatology offers a valuable lens through which to understand a broader eschatological hope that encompasses the entire created order. By examining this shift, I aim to demonstrate how Pentecostalism's evolving theological orientation can inform a pneumatological vision of animal eschatology. While Pentecostal theology traditionally centred on human salvation, recent movements towards a more creation-inclusive eschatology reflect a growing recognition of the Spirit's redemptive work across all of creation, opening new avenues for ecological and animal theology.

Here, I propose that a Pentecostal reimagined eschatology, which increasingly emphasises continuity and transformation over apocalyptic destruction, provides a framework for integrating animals into this eschatological vision. This inclusion is not incidental but central to a holistic understanding of salvation and restoration that values animals as part of God's beloved creation. In bringing animals into this framework, I seek to expand Pentecostal eschatology's potential to contribute to contemporary discussions on creation care and to position animals within the future hope that this tradition offers.

Within Pentecostalism, the call for a creation-inclusive eschatology is particularly pertinent. Historically, Pentecostals have tended to be slower to engage with ecological issues, primarily due to dispensationalist eschatology's emphasis on an apocalyptic end of the world.[7] This eschatological orientation often results in a diminished regard for environmental stewardship, as the anticipation of an imminent end discourages any perceived need for long-term ecological responsibility. Scholars like Swoboda and Althouse argue for a "Green Pentecostal Pneumatology," suggesting that the Spirit's indwelling presence within creation implies a divine mandate to care for the environment (including animal life).[8]

[7] Newton, "Apocalyptic Ecotheology: A Pentecostal Contribution."

[8] Newton, "Apocalyptic Ecotheology," (pp) *forthcoming*. See also Swoboda, *Tongues and Trees*; Swoboda, "Reconciling Creation"; Althouse, "Implications of the Kenosis

This approach reframes Pentecostal eschatology as not only anticipating the redemption of human souls but also the restoration of the entire created order. Newton discusses how Revelation's ecological imagery, particularly its portrayal of environmental destruction and renewal, can inform this approach. Newton highlights that the biblical text of Revelation underscores this creation-centred vision, standing alongside Genesis as one of the most environmentally conscious books of the Bible. Revelation's language of creation worshipping God, its vivid portrayals of environmental crises, and its warnings to those who "destroy the earth" (Rev 11:18) collectively form a compelling basis for ecological responsibility. Alan Cadwallader states that Revelation 22:1–5 presents a vision "of Earth's creatures living in life-affirming ecological interdependence."[9] In this way, Revelation serves as a significant resource for envisioning an eschatological future where all creation is renewed and sustained under God's justice and mercy.[10] Thus, Revelation offers a compelling theological basis for a future in which all creation is renewed and sustained, reinforcing a call to ecological responsibility.

This reinterpreted Pentecostal eschatology, which emphasises transformation rather than annihilation, holds significant implications for animal pneumatology and eschatology. By viewing ecological stewardship as part of the Spirit's mission, Pentecostal theology expands its eschatological vision to encompass all of creation, including animals.

This reimagining of eschatology aligns with a broader shift in Pentecostal theology, as described by feminist scholars such as Davis and Grey. In recent decades, Pentecostal theology has matured, prompting critical reassessment of its eschatological focus, particularly as it moves beyond the doctrinal constraints of premillennial dispensationalism. Davis and Grey note that the traditional premillennial perspective, which emphasises the imminent destruction of creation prior to Christ's return, has historically impeded the integration of ecological concerns within Pentecostal thought.[11] Originally formulated in response to debates over origins and eschatological endpoints in the late 19th and early twentieth centuries,

of the Spirit for a Creational Eschatology: A Pentecostal Engagement with Jürgen Moltmann."

[9] Cadwallader, "Earth as Host or Stranger?," 170.

[10] Newton, "Apocalyptic Ecotheology: A Pentecostal Contribution."

[11] Davis and Grey, "The Apocalyptic Destruction of the Earth in Isaiah 24 and Its Place in a Pentecostal Ecotheology."

this doctrine's apocalyptic focus limited the theological space for ecological engagement, framing environmental issues as ultimately insignificant in light of an impending apocalyptic end.[12] This response critiques the dispensationalist hermeneutical framework and its disconnect from the lived Pentecostal experience.[13]

Instead, Pentecostal thought has increasingly embraced a proleptic eschatology, which recognises that the final realisation of the new creation has already been inaugurated in the resurrection and ascension of Christ. Consequently, most Pentecostals believe they now live within the tension of the "already" and "not yet," experiencing the kingdom of God through the outpoured Spirit while also awaiting its ultimate fulfilment when God will be "all in all." Jesus Christ, as the initiator of this kingdom, embodies the divine reign of love on earth, even as believers anticipate the final consummation of God's will.[14] This proleptic vision echoes early Pentecostal fervour for the "latter-day outpouring" of the Spirit and the anticipation of Christ's return. Davis and Grey note Macchia's emphasis on the timeline of this ultimate fulfilment, which remains known only to God, as outlined in 2 Peter 3:8–10. This shift in eschatological thinking reorients Pentecostalism towards a view of new creation not as annihilation but as transformation. The new creation is not *ex nihilo* (out of nothing) but *ex vetera* (out of the old). The outpoured Spirit, the Spirit of life, is already at work, bringing about the eschatological renewal of creation in the present moment. Steven Land's apocalyptic eschatology, for instance, reimagines the "last days" as the in-breaking of the Spirit's work in creation rather than the world's destruction.[15]

This vision affirms that the entire world, not just humanity, is within the scope of the Spirit's redemptive mission. The resurrection of the body further highlights the value of the material world and God's intention to preserve and restore creation, rejecting any eschatological escapism that dismisses ecological responsibility. Pentecostal eschatology, then, is one of hope and renewal, grounded in Christ and the event of Pentecost, rather than in a destructive apocalypticism devoid of hope for creation. The "former things" that pass away (Rev 21:4) are not the material world itself

[12] Norris, "Creation Revealed," 80–82.

[13] Yong, *In the Days of Caesar: Pentecostalism and Political Theology*, 316–58.

[14] Davis and Grey, "The Apocalyptic Destruction."

[15] Yong, *In the Days of Caesar*, 331.

but the forces of death, grief, and suffering that weigh upon creation, which, as Rom 8:20–21 attests, groans for liberation.[16]

Understood in this light, the Spirit poured out at Pentecost is not merely a signal of personal transformation but the eschatological force that catalyses the renewal of all creation, human and non-human alike. The hope of transformation extends beyond the individual soul to encompass social structures and the entirety of the created order. This transformative work, inaugurated in Christ and anticipated at Pentecost, is prefigured through the faithfulness of believers who live under the lordship of Christ, signalling the "already" of the kingdom even as they await the "not yet" of its ultimate fulfilment. Charismatic gifts, such as healing, both in the ministry of Christ and within the life of the church, serve as tangible signs of this redemptive trajectory, affirming not just the spiritual but the material renewal of all things.[17] This theological vision offers a decisive counterpoint to the dualistic body-soul framework found in premillennial dispensationalism, which tends to devalue the material world in favour of a purely spiritual destiny. Instead, Pentecostal eschatology, rooted in the pneumatological work of the Spirit, embraces a holistic view of redemption that honours the materiality of creation and affirms its ultimate restoration. In this light, the mission of the church is broadened: it is not solely concerned with individual salvation but includes active, faithful engagement with the world. Such engagement serves as a witness to God's transformative work, revealing that the renewal of creation is already underway and will be consummated in the eschaton. The Spirit of Pentecost, then, empowers believers not only for personal sanctification but also for participation in the divine mission to renew the world, thereby affirming the intrinsic worth of creation itself.

This renewed Pentecostal eschatology, with its emphasis on the Spirit's role in cosmic renewal, logically extends to encompass animals as well. In this framework, animals are not peripheral to God's redemptive plan but integral to the restoration of all creation. As Pentecostals

[16] Peter Althouse explores these themes in depth in his chapter on the eschatological dialogue between Moltmann and Pentecostalism, where he examines how Moltmann's cosmic vision resonates with and challenges Pentecostal eschatological perspectives. Peter Althouse, Spirit of the Last Days: Pentecostal Eschatology in Conversation with Jurgen Moltmann, *Journal of Pentecostal Theology*: Suppliment Series 25 (Edinburgh: T&T Clark, 2003), 158–92.

[17] Rizzo, "Pentecostal and Evangelical Ecotheologies."

shift from an apocalyptic vision of annihilation to one of transformative renewal, they open theological space for understanding animals as participants in, and beneficiaries of, the Spirit's work. Thus, the eschatological hope offered through Pentecostal thought is one that includes the flourishing of all creatures, acknowledging animals as fellow recipients of the Spirit's sustaining presence and ultimate redemption.

I propose that an additional dimension of Pentecostal eschatology, rooted in revival traditions, includes practices like snake handling and the belief that faith protects believers from snake bites.[18] The Hebrew Scriptures affirm that one of the blessings of wisdom is peace with wild animals, as seen in Job 5:22–23. This concept may inform the New Testament portrayal of snake handling as a demonstration of faith (Mark 16:16; Luke 10:19). In handling one of nature's most dangerous creatures as if it were tame, believers symbolically step into the eschatological vision of harmony in creation, a vision that reverses the conflict introduced in the Genesis narrative.[19] This act reflects a biblical ideal of nature as relational and harmonious, embodying an effort to reconcile with the creature traditionally seen as humanity's adversary in the Genesis story.

The prophetic literature, particularly in Isaiah, envisions a Spirit-empowered Messiah whose coming transforms not only human relationships but the entirety of creation. Pentecostal theology, with its robust emphasis on the Spirit's pervasive and transformative activity, resonates deeply with this vision. This messianic hope gestures towards a future

[18] For a historical background on Pentecostal snake handling practices, see Kamenicky, " 'You Will Not Surely Die,'" 191–94. and Reed, "Serpent Handling Pentecostal Churches: Revisiting Disputed Borders." While I do not endorse the literal practice of snake handling as observed in some fundamentalist traditions, my focus here is on its potential symbolic significance. In my view, traditional snake handling is primarily practiced as a demonstration of faith or divine protection rather than an act of reconciliation with animals. However, symbolically reinterpreted, it could be seen as a gesture towards an eschatological hope for restored harmony in creation, a vision in which humans and animals coexist peacefully, reversing the conflict depicted in the Genesis narrative. This reimagined symbolism aligns with Pentecostal eschatology's potential to embrace a holistic view of creation's renewal. Andrew Linzey would condemn this practice, arguing that handling animals as symbols of faith rather than recognising their value and autonomy risks reducing them to mere tools for human purposes. In contrast, a symbolic interpretation that honours animal rights would resonate with an ethical vision of creation's restoration, one where all creatures are respected as co-participants in God's creation. I lean on the practical ethics of his book, Linzey, *Why Animal Suffering Matters*.

[19] Webb, *On God and Dogs*, 171.

where justice, peace, and righteousness saturate all creation, affirming that the Spirit's work is not confined to human renewal but is inherently inclusive, permeating the fullness of the created order. Such an approach provides a dynamic lens through which Pentecostal pneumatology might continue to address the broader implications of the Spirit's redemptive activity for animals.

Anita Davis and Jacqueline Grey's interpretation of Isaiah 24 contributes to a pneumatological paradigm that enriches an animal-inclusive eschatology, aligning with themes in Moltmann's theology. They highlight how the apocalyptic vision in Isaiah 24, rather than predicting total destruction, points towards a cosmic judgement that initiates transformation.[20] They frame the Spirit as an active, renewing presence within creation rather than a force solely focused on human redemption. In Isaiah, the Spirit's influence extends across the entire breadth of creation, suggesting that all creatures participate in the eschatological hope of renewal.[21] This universal covenant, rooted in the Noahic covenant's promise to "all flesh," also implies a shared destiny for animals and humans within God's redemptive plan.[22] Here, I reflect that amid

[20] Davis and Grey, "The Apocalyptic Destruction of the Earth."

[21] Hausoul raises a thought-provoking question regarding the fate of animals in the eschatological vision of the new heavens and new earth: will new animals be created, or will existing animals be resurrected? Many Christians lean toward the idea of God creating new animals, as it can be difficult to envision every creature that has ever lived being restored. However, this raises a relational dilemma, our bonds are with specific animals we have known, like a beloved pet, rather than with an abstract "new" version of a cat or dog. Hausoul suggests another possibility: in the renewed creation, animals might live free from death, allowing those cherished relationships to endure within the transformed creation. Hausoul, *God's Future for Animals*, 201. Ryan Patrick McLaughlin further expands on this issue by addressing the challenge of representation in eschatology. He argues that the resurrection of all individual animals preserves the continuity of species within eschatological life. McLaughlin contends that creating a single "representative" of each species would fail to account for the individuality and unique histories of each creature, thereby undermining the meaningfulness of their eschatological inclusion. The resurrection of all creatures, rather than a mere symbolic representative, honours the particularity of each life, aligning with a vision of cosmic redemption that upholds both individuality and collective identity. McLaughlin, "Possible Critiques of Cosmocentric Transfiguration," 376–77. This question will be further explored in the coming sections with Moltmann's insights.

[22] For David Clough, the Noahic covenant demonstrates how "it is not only human animals that are addressed by God and called to live lives in response to God." Clough, *On Animals, Systematic Theology*, 41.

destructive forces, a remnant of animal life is saved alongside the human. The value of animal life is emphasised not just through preservation but also through promise (covenant).

Seen through this lens, Davis and Grey sketch a vision in which the Spirit gathers her non-human creatures into the slow renewal of the world. Isaiah 24's apocalyptic imagery lays bare humanity's responsibility for ecological degradation yet also suggests a covenantal framework in which creation's brokenness is met with divine restoration. This pneumatology, therefore, sees animals as integral to the Spirit's renewing activity and ultimate reconciliation.

This reading of Isaiah 24 contributes to an animal eschatology in which the Spirit's life-giving and sustaining work extends to all creatures. This vision aligns with a pneumatology that anticipates the transformation of creation itself, where animals, like humans, are included in God's covenantal promise and share in the eschatological hope. The Spirit's work is, therefore, not limited to human redemption but embodies a broader ecological and creaturely renewal, positioning animals within a narrative of eschatological reconciliation.

Eschatology as Reorientation

Ernst Conradie offers a vital corrective to eschatological models that centre on human destiny while relegating the rest of creation to the margins. His eschatology deliberately widens the frame, embracing animals and the broader natural world within the Spirit's transformative mission. What is compelling here is not simply Conradie's inclusion of non-human creatures, but his reconfiguration of eschatology itself. Rather than treating eschatology as a nostalgic return to some primordial ideal, Conradie envisions it as an ongoing process of sanctification and glorification—an unfolding transformation in which the Spirit draws creation, in all its fractured materiality, towards its divinely intended future.[23]

[23] Ernst M. Conradie, "What Is the Place of the Earth in God's Economy? Doing Justice to Creation, Salvation and Consummation," in *Christian Faith and the Earth: Current Paths and Emerging Horizons in Ecotheology*, ed. Ernst M. Conradie (New York: Bloomsbury T&T Clark, 2014), 69. See also, Ernst Conradie, *The Earth in God's Economy: Creation, Salvation and Consummation in Ecological Perspective* (Zürich: Lit Verlag, 2015), 274–87.

I find this move particularly significant for animal theology. Romanticised "Edenic" visions of eschatology often risk flattening creation's present complexities, treating suffering and ecological degradation as aberrations to be erased rather than realities through which God is actively working. Conradie resists this tendency. He insists that eschatology must grapple with the material, relational, and spiritual intricacies of the world as it is, not merely as we wish it to be. In doing so, he opens theological space to recognise animal suffering and ecological disruption as part of the terrain through which divine redemption moves, not simply as obstacles to be reversed.

At the heart of Conradie's vision lies a paradox: creation is both beloved and in need of transformation.[24] It is already deemed "worth dying for," yet remains incomplete, yearning towards a future where divine love is fully realised. This dual posture—love and transformation held together—guards against two opposite errors: romanticising the past and escaping the present. Eschatology, for Conradie, is neither retreat nor flight; it is the Spirit's dynamic reorientation of all things towards God's purposes.

This reorientation moves into continual movement from inherited frameworks (political, cultural, and religious) towards perceiving the world "in the light of the Light of the world."[25] Liturgy and gospel narrative function here not merely as devotional practices, but as interpretive lenses that teach us to see creation anew: as beloved, complex, and destined for transfiguration. This eschatology of redefinition refuses to

Conradie posits human sin as the fundamental issue requiring resolution. The incarnation, in this context, is perceived as a response solely to the fall of humanity, serving to restore the fractured relationship between the Creator and his creatures. Further, Conradie interprets Arnold van Ruler's view of the incarnation as an 'emergency measure,' which implies that sin was not an inevitable aspect of creation. Accordingly, salvation is understood not as a remedy for a fundamental flaw but as a process of restoring all creatures to a state of harmony within God's presence. Within the Reformed tradition, this idea is often described as 're-creation,' emphasising the eschatological aim of God's salvific acts. This 're-creation' represents a transformative work, focused on healing creation and realigning it with God's original intent and ultimate purpose. Conradie, "What Is the Place of the Earth in God's Economy?," 85; Van Ruler, *Calvinist Trinitarianism and Theocentric Politics: Essays Toward a Public Theology*, 38:131.

[24] Conradie, "What Is the Place of the Earth in God's Economy? Doing Justice to Creation, Salvation and Consummation," 94.

[25] Conradie, 94.

reduce salvation to human ascent or cosmic erasure.[26] Instead, it imagines the cosmos—material, spiritual, relational, and animal—as being lovingly redrawn into God's eschatological communion.

Conradie's approach resonates with Indigenous eschatological perspectives, particularly Native American theological visions in which the "end" is not a cataclysmic finale but the restoration of balance within creation.[27] Here, ecological degradation is more than a physical problem; it signals spiritual dislocation, a rupture in the fabric of relational harmony between humans, animals, land, and Creator. Such a vision enriches Conradie's theological reorientation by grounding it in relationality rather than linear progression. Hope is found not in annihilation or escape but in the Spirit's work of drawing all creatures back into right relationship—humans with animals, communities with ecosystems, creation with its Creator.

By bringing Conradie's eschatological imagination into dialogue with Indigenous perspectives, a more expansive pneumatological horizon emerges. This horizon challenges both escapist theologies that seek heaven apart from creation and reductive visions that fail to reckon with creation's groaning. Instead, it invites us to see the Spirit's work as both transformative and deeply attuned to the creaturely and ecological realities of the present world.

Eschatology as Divine Justice

Jürgen Moltmann's eschatological vision is one of the most fertile theological frameworks for reimagining humanity's relationship to the rest of creation.[28] What makes Moltmann so compelling for animal theology is

[26] Conradie's vision promotes an eschatology in which God's saving work involves all levels of complexity, from the "biophysical" to the highest dimensions of consciousness and spirit. This nuanced view of eschatology seeks to avoid escapism and recognises that the material world, with all its depth and diversity, has a role in God's redemptive plan that is deeply integrated with the spiritual. Thus, eschatology, for Conradie, is about the holistic transformation and glorification of creation as God's beloved rather than its mere replacement, elevation, or destruction.

[27] Kidwell, Noley, and Tinker, *A Native American Theology*, 152–55. This view challenges eschatologies that imagine a complete obliteration of the current world, instead offering a vision where the "end" is marked by a re-balancing or reorientation toward life-sustaining harmony.

[28] I draw heavily on Ryan Patrick McLaughlin's interpretations of Moltmann's eschatology, particularly his insights into the theological implications of creation's renewal. McLaughlin's work provides a robust foundation for understanding Moltmann's vision

his refusal to treat non-human life as a peripheral concern. For him, the worth of every creature (human and non-human alike) is not a theological afterthought but a central, animating conviction. Grounded in divine love and sustained by the life-giving work of the Spirit, every creature is affirmed in its dignity and right to exist as part of the broader community of creation. Moltmann's insistence that animals are not merely instrumental to human flourishing but participants in creation's shared destiny forces a reorientation of eschatological imagination itself.

Moltmann thus seeks to move beyond a narrow, human-centred view of redemption and consider a reconciliation that embraces the entirety of creation. He pushes for a theological imagination that perceives human responsibility to the earth and its creatures not simply as a moral obligation but as an extension of our shared existence within creation. Justice and peace, in his vision, cannot remain human enclaves; they must extend across the whole spectrum of life. For Moltmann, the call to recognise animal rights emerges from a theological conviction that animals, like humans, are inherently valuable as part of God's creation.[29] This value

of eschatological harmony, which I build upon in this discussion. McLaughlin's work has been invaluable in exploring eschatological themes with a strong engagement with the Spirit, particularly within the context of animal theology. His nuanced treatment of Moltmann and Linzey provides essential anchor points for understanding the role of the Spirit in the eschatological renewal of all creation. McLaughlin's insights have been instrumental in shaping this discussion, offering a foundation for envisioning a future where animals and the entirety of creation participate in divine communion. Particularly his texts: Ryan Patrick McLaughlin, "Anticipating a Maximally Inclusive Eschaton: Jürgen Moltmann's Potential Contribution to Animal Theology," *Journal of Animal Ethics* 4, no. 1 (October 24, 2020): 18–36; McLaughlin, "Jürgen Moltmann (1926–): Creation and Sabbath Theology," in *Animal Theologians*, ed. Andrew Linzey and Clair Linzey (New York: Oxford University Press, 2023), 401–21.

Jürgen Moltmann, *The Coming of God: Christian Eschatology*, trans. Margaret Kohl (Minneapolis, MN: Fortress Press, 1996), 259.

[29] This perspective aligns with a radical assertion that animals possess rights deserving of legal protection, suggesting that contemporary society must move beyond exploitative practices to honour the unique nature of each creature. By referencing biblical and historical precedents, such as Leviticus 20:15 and medieval legal codes, Moltmann illustrates that past societies, though imperfectly, acknowledged a need to balance human and animal interests within legal systems. In Moltmann's view, animals should not be relegated to positions of subjugation or exploitation, as evident in his critique of institutional settings like zoos, which he sees as prisons for animals. His theological framework calls for a sanctified lifestyle that actively defends animal rights against human aggression. For Moltmann, the ultimate reconciliation of creation entails a peaceable community underpinned by justice and legal protections that sustain both human and non-human life,

is not derived from their utility to humans but from their status as beings beloved by God.

Moltmann's panentheistic cosmology deepens this claim. Throughout his extensive body of work, Moltmann develops a theological foundation for an eschatology that encompasses all creatures.[30] Eternal life, in Moltmann's thought, is understood as a creaturely existence that transcends mortality through enduring participation in the divine source of life. God, who is both immanent and transcendent, indwells all of creation, both in heaven and on earth, sharing with every creature a portion of the divine vitality.[31] This panentheistic perspective envisions creation and the Trinity as mutually interwoven, with the world existing within the divine life even as it maintains its own distinct identity. Creation, in Moltmann's words, becomes "God's eternal home country," while God, in turn, becomes "the eternal home of everything he has created."[32] I find this particularly significant because it relocates animals and ecosystems not to the edges of eschatological hope but to its centre. Every instance of existence, every individual life, is gathered up into this eschatological horizon.

aligning human actions with God's intention for harmony across the cosmos. Moltmann, *Creating a Just Future*, 69–71; Moltmann, *The Source of Life*, 50; Moltmann, *The Spirit of Life*, 36; Jürgen Moltmann, *The Spirit of Hope: Theology for a World in Peril*, trans. Margaret Kohl (Louisville, KY: Westminster John Knox Press, 2019), 87. Moltmann goes further to suggest that nature (which incudes the nonhuman animal) possesses the right to exist for its own sake. Jürgen Moltmann, *Ethics of Hope* (Minneapolis, MN: Fortress Press, 2012), 84.

[30] Central to his framework is the concept of the Trinity as not only social but expansively open. Moltmann describes the Trinity as a dynamic and inclusive community, where "the trinitarian relationship of the Father, the Son, and the Holy Spirit is so wide that the whole creation can find space, time, and freedom in it" Moltmann, *The Spirit of Life*, 109. 10/02/2026 12:46:00. This portrayal of the Trinity as inclusive extends to all of creation, aligning with Moltmann's stance against divine impassibility, which he sees as incompatible with the deeply involved and loving nature of God. The crucifixion, for Moltmann, serves as the ultimate manifestation of this love, bridging the divine and the created world. So, McLaughlin, "Anticipating a Maximally Inclusive Eschaton," 20. The Trinitarian community, in Moltmann's view, is thus characterised by a love that transcends its own boundaries, actively seeking communion with 'others.' This raises the question of who or what constitutes the "other" in this relationship.

[31] Jürgen Moltmann, *The Way of Jesus Christ: Christologyin Messianic Dimensions*, trans. Margaret Kohl (Minneapolis, MN: Fortress Press, 1993), 333.

[32] Jürgen Moltmann, *In The End-The Beginning: The Life of Hope*, trans. Margaret Kohl (Minneapolis, MN: Fortress Press, 2004), 157–58.

This vision of divine and cosmic interdependence advances an inclusive eschatology that anticipates the resurrection of all forms of life. Each individual life, each instance of existence throughout history, is embraced within this divine reality, leading to a vision of resurrection that extends beyond humanity alone.[33] In this way, Moltmann's theology articulates a hope for the ultimate redemption and renewal of the entire cosmos, where the inherent value of each creature is affirmed, not only within the present creation but also in the eschatological fulfilment of all things.[34] This eschatological lens invites a reimagining of creation as a community destined for participation in the life of God, a vision of universal reconciliation in which all creatures find their place within the unfolding drama of redemption.

In Moltmann's theological lens, the incarnation and resurrection are foundational events that uphold the Christian promise of hope and transformation, casting a redemptive light that reaches far beyond humanity alone. This Christological inclusivity suggests that Christ's life, death, and resurrection are not restricted to human redemption but encompass all of creation in their scope.[35] Moltmann articulates this universal reach by asserting that "Christ died the death of all the living in order to reconcile them all," a claim that situates every creature, human and non-human alike, as recipients of Christ's redemptive act.[36] This inclusivity ascribes inherent worth to all of creation, affirming that each creature holds a divine right to life, grounded in Christ's sacrificial love. For Moltmann, Christ's redemptive work is incomplete if it does not embrace the entirety of the cosmos; thus, the reconciliation of all creation is essential for the fullness of Christ's identity and mission. This comprehensive approach to redemption underscores a deep reverence for creaturely life in all its forms, urging an eschatological vision that includes the suffering and vulnerability of every creature within the broader redemptive arc.

What I find theologically generative here is Moltmann's treatment of incarnation and resurrection as events of cosmic solidarity. Christ becomes the "ultimate victim" of evolutionary pain and death, gathering

[33] Moltmann, *God in Creation (The Gifford Lectures, 1984–1985)*, 258.

[34] Moltmann, *The Spirit of Life*, 219.

[35] Moltmann, *The Way of Jesus Christ*, 152–53.

[36] Moltmann, *The Coming of God*, 92–93.

all creaturely pain into the Trinitarian life through his resurrection.[37] As Moltmann states, "unless the whole cosmos is reconciled, Christ cannot be the Christ of God and cannot be the foundation of all things."[38] This framing does more than expand the scope of redemption; it reframes suffering itself. Animal suffering is not ignored, minimised, or absorbed into human narratives. It is gathered, honoured, and ultimately transfigured in Christ. Therefore, Moltmann's Christology provides a robust theological framework for universal reconciliation, a vision where all of creation, down to the smallest and most vulnerable forms of life, is drawn into the redemptive embrace of Christ's incarnational and resurrected presence.

This is a deeply pneumatological vision as well, for it is the Spirit who mediates this gathering, sustaining creation in its groaning and orienting it towards new creation, as Ryan Patrick McLaughlin highlights this creaturely inclusion.[39] The Spirit is the breath animating all life, the presence resonating in creation's sighs and groans, and the eschatological guide leading all things towards their consummation. These three dimensions—animating, sustaining, and transforming—form a pneumatological arc that stretches across the entirety of creaturely existence.

Importantly, animals are encompassed as part of the "all flesh" that participates in the Spirit's redemptive activity. Such a vision grants dignity and worth to non-human creatures, suggesting that they, too, are recipients of the Spirit's work and integral participants within God's creative and redemptive purpose.[40] Moltmann's pneumatology thus opens theological space to affirm animals as beings of value in themselves, not merely as props to human redemption, but as part of the eschatological renewal.[41] For Moltmann, the eschatological scope of redemption

[37] Moltmann, *The Way of Jesus Christ*, 153; Moltmann, *In The End-The Beginning*, 70.

[38] Moltmann, *The Way of Jesus Christ*, 306.

[39] McLaughlin, "Anticipating a Maximally Inclusive Eschaton," 25.

[40] Moltmann, *The Way of Jesus Christ*, 256.

[41] Moltmann's vision of creation as a coherent, eschatologically oriented process offers a vital framework for understanding the Spirit's work among all living beings. By magnifying the unity of God's creative activity across time, Moltmann affirms that animals, alongside humanity, are integral to the unfolding of creation's ultimate purpose. This challenges the human-centric tendency to view animals as secondary and instead situates them as participants in the Spirit's redemptive movement. Moltmann, *The Future of Creation*, 119.

includes all of creation, such that "all flesh" will experience resurrection and redemption, extending beyond humanity to embrace non-human animals.[42] Each creature's existence, from its past experiences to its future hopes, is held within the arc of divine redemption, woven into the larger narrative of the Spirit's work in creation. This universal resurrection does not merely provide a continuation of life but envisions the completion and fulfilment of each creature's life in communion with God.[43] Moltmann's theology, then, does not merely allow for animal inclusion; it necessitates it.

In Moltmann's eschatological vision, the resurrection of nature is not to be dismissed as mere sentimentality or romantic idealism; rather, it is a profound expression of divine justice.[44] The fullness of God's justice demands a universal scope of redemption that encompasses not only humanity but the entirety of creation.[45] To exclude any creature from the arc of redemption would be to render God's justice incomplete, leaving the divine promise unfulfilled. Thus, Moltmann insists, "If we were to surrender hope for as much as one single creature, for us God would not be God" – any limitation in the reach of resurrection would undermine the very character of God's justice and love.[46]

[42] Moltmann, *Ethics of Hope*, 84.

[43] Moltmann, *The Way of Jesus Christ*, 256; 333. As Moltmann writes, the Creator's Spirit breathes life into all creatures, uniting them through God's sustaining power. The Spirit, experienced as both a life-giving presence and as an open "space" where life unfolds, allows for a conception of God that is not merely personal but also profoundly spatial and communal. This understanding aligns with traditions that experience the Spirit as *Ruach*, God as the very breath that enlivens and holds all creatures within a shared, divine embrace.

[44] Moltmann, *Ethics of Hope*, 84.

[45] Moltmann emphasises that justice is the foundation of authentic independence—between individuals, societies, and humanity's relationship with the environment. Justice, as he describes it, arises from the symbiosis of diverse creaturely life grounded in the mutual recognition of their independence and subjectivity. This principle not only undergirds shared survival but also aligns with an eschatological doctrine of creation. In the context of animal pneumatology, this framework underscores the Spirit's role in fostering interdependent relationships among all life forms, including animals. Recognising and respecting the value of animals becomes vital to envisioning a Spirit-filled creation where justice and mutual flourishing extend across species. Without embracing this vision, humanity's survival remains inseparable from the fate of non-human creatures. Moltmann, *The Future of Creation*, 130.

[46] Moltmann, *The Coming of God*, 132.

This inclusive vision is grounded in Moltmann's understanding of covenant, wherein God's promise is established not only with humanity but with every living creature. Here, covenantal inclusivity reshapes the theological landscape—a reminder that humans are not the sole participants in the divine dialogue. All creatures are summoned into relationship with God, affirming a solidarity that spans species within the divine covenantal order. The Spirit, present in the risen Christ, actively intercedes for all of creation, upholding its destiny in the face of mortality and decay. Thus, even as creation groans under the weight of corruption, the Spirit sustains its future within God's redemptive embrace, ensuring that all forms of life are drawn into the ultimate reconciliation that is both promised and secured by God.

In his chapter, "Theological Foundations for Cosmocentric Transfiguration," McLaughlin sees eschatology as a process whereby God brings everything He created, humans, animals, and the whole cosmos, into final, perfected communion with Himself.[47] This future includes transformation rather than destruction or replacement and suggests that God's ultimate aim is a world where no part of creation is isolated or excluded from divine love and relationship.

McLaughlin provides a nuanced exploration of eschatology that intertwines creation, communion, and consecration within the theological frameworks of Jürgen Moltmann and Andrew Linzey. McLaughlin argues that all of God's creation is consecrated through a process of "separation", whereby the world is made distinct from God, enabling it to become "the other-with-God" in communion. However, this separation carries inherent risks, manifesting as isolation, thus necessitating restoration. For McLaughlin, both Christ and the Spirit initiate this restorative process, though they do not fully accomplish it in historical time. The cosmos, having adapted to its state of isolation, remains entrapped within evolutionary mechanisms that perpetuate suffering and mortality. This is acknowledged by both Moltmann and Linzey, who recognise the limitations of the present creation and the need for an eschatological transformation to overcome these conditions.

McLaughlin interprets Moltmann's eschatology as centring on judgement, a divine act of consecration that restores creation and ultimately

[47] Ryan Patrick McLaughlin, *Preservation and Protest Theological Foundations for an Eco-Eschatological Ethics*, Emerging Scholars (Minneapolis, MN: Fortress Press, 2014), 339–41.

abolishes death and suffering. This ultimate communion involves transfiguration rather than replacement; it is the transformation of the present creation, retaining lines of continuity with its current form but elevated into a state of perfect communion with God. This vision entails a "new creation" that is not numerically different but fundamentally transfigured, similar to the resurrection event, which exhibits both continuity and discontinuity with its pre-resurrection form, as noted by theologian Celia Deane-Drummond.[48]

From McLaughlin's perspective, resonating with Karl Barth's christological focus, eschatology represents the completion of Christ's work. It constitutes not merely an "end" to history but God's full arrival, the culmination of divine communion that must ultimately encompass all of creation. McLaughlin contends that this eschatological vision cannot be fulfilled unless all creatures are included within it; otherwise, God's original intent for creation remains perpetually unrealised. He argues, following Keith Ward, that "immortality for animals as well as humans is a necessary condition of any acceptable theodicy."[49] This inclusive redemption requires the resurrection and transfiguration of every aspect of creation, sentient and non-sentient alike, to fulfil God's desire for communion with all creation.

In sum, McLaughlin's engagement with Moltmann and Linzey proposes an eschatology in which all creation, by virtue of its consecration, is ultimately drawn into divine communion. This vision challenges anthropocentric theologies by affirming that every creature, possessing its own narrative within the broader cosmic story, is swept into the divine narrative of redemption. The eschaton, understood as "God's coming and arrival,"[50] signals a fulfilment of God's desire for communion, in which every element of creation participates in the redemptive process that Christ initiated and the Spirit continues to sustain.

[48] McLaughlin, 341. In a similar vein, Peter Althouse emphasises that Moltmann's cosmic eschatology insists on a transformed world rather than one destined for destruction. This vision resists the notion of an apocalyptic eschatology associated with nuclear or ecological catastrophe, which Moltmann argues is both theologically flawed and unnecessarily alarming. Instead, true apocalyptic, rightly understood, signifies the eradication of sin and evil rather than a cataclysmic end of the world. Althouse, *Spirit of the Last Days*, 116–17.

[49] McLaughlin, *Preservation and Protest* 340.

[50] McLaughlin attributes this phrase to Moltmann. McLaughlin, 340.

Future-Oriented Animal Eschatology

Celia Deane-Drummond, in her critique of Moltmann's approach to the Spirit's role in creation, commends his holistic vision while pointing out critical gaps in detail. She argues that Moltmann's expansive view risks over-generalisation, portraying the Spirit's work as universally inclusive without adequately distinguishing how the Spirit might uniquely engage with non-human creation.[51] Specifically, Deane-Drummond highlights that Moltmann does not offer a concrete framework, or *Gestalt*, for understanding the Spirit's relationship with animals, nor does he address how the Spirit might interact with non-human life in practical, future-oriented ways.[52]

In response, I aim to build on Deane-Drummond's critique by proposing that the Spirit's work encompasses specific, future-oriented ways that envision concrete actions or outcomes aimed at the flourishing, restoration, and transformation of non-human creation within an eschatological framework. This includes understanding the Spirit as an advocate, grieving over the extinction and suffering of animals; as a healer, fostering restoration in ecosystems and animal well-being; and as an agent of renewal, working towards a creation-wide transformation that begins in the present while holding in tension the eschatological "not yet." By proposing these future-oriented ways, I seek to address Deane-Drummond's critique directly, and to envision the Spirit's work as actively supporting animal flourishing not only in abstract theological terms but through tangible outcomes that align with the Spirit's mission of holistic redemption. This approach advances that animals are not peripheral to

[51] Celia Deane-Drummond, *Ecology in Jürgen Moltmann's Theology* (Eugene, OR: Wipf & Stock, 2016), 233. For example, John Polkinghorne raises an important critique of Moltmann's eschatology, suggesting that Moltmann's expansive view of resurrection, wherein 'resurrection has become the universal "law" of creation, not merely for human beings, but for animals, plants, stones, and all cosmic systems as well', risks turning the eschaton into a static 'museum collection' of everything that has ever existed. While Moltmann's approach seeks to affirm the value and destiny of all creation, Polkinghorne questions the necessity of preserving every individual element (such as individual stones) in an eschatological framework, thereby highlighting a tension between universal preservation and meaningful eschatological transformation. John Polkinghorne, *The God of Hope and the End of the World* (New Haven, CT: Yale University Press, 2002), 123, http://ebookcentral.proquest.com/lib/harvest/detail.action?docID=3420227.

[52] For further engagement with critiques on Moltmann's cosmology, see McLaughlin, "Possible Critiques of Cosmocentric Transfiguration."

eschatological hope but are intrinsic participants in the Spirit's redemptive and life-giving mission.

Spirit as Advocate for All Creation

Moltmann's eschatological vision provides a powerful starting point for reimagining the Spirit's role as advocate within a suffering creation. His theology, rooted in divine justice, refuses to confine redemption to humanity alone. Instead, he anticipates a future in which the suffering and decay that weigh upon all living beings are transfigured into the freedom and life of God's Kingdom (Rom 8:19–22). What is striking here is Moltmann's insistence that eschatology is not merely a distant horizon but an active divine engagement that confronts and dismantles systems of oppression in the present. This theological posture compels a broader view of redemption—one that refuses to ignore animal suffering as an incidental byproduct of human sin. Animals, too, are caught within exploitative structures and are thus integral to the Spirit's eschatological mission of liberation and renewal.

Building on this foundation of eschatological justice, Mark Wallace's pneumatology offers a complementary view of the Spirit as a "comforter," "advocate," and "healer" on behalf of all creation.[53] If the Spirit stands with the marginalised and the oppressed, then this divine solidarity extends naturally to animals, who are often voiceless victims of human exploitation through practices like industrial farming, habitat destruction, and the impacts of climate change. Wallace's depiction of the Spirit as an advocate for the "silenced" thus expands the theological understanding of the Spirit's role to include all marginalised beings, human and non-human alike. This advocacy empowers a theological recognition of animals' dignity and worth, suggesting that they, too, are recipients of the Spirit's comforting and healing presence.[54]

[53] Mark I. Wallace, *Fragments of the Spirit: Nature, Violence and the Renewal of Creation* (New York: Continuum Publishing, 1996), 212.

[54] The comforting power of animals in moments of human grief and distress reflects a reciprocal relationship often mediated by the Spirit. Just as humans may experience comfort and healing through the companionship of domestic animals, these creatures, animated by the Spirit's life-giving presence, may also share in this mutual exchange of care and connection. This dynamic suggests that animals can be active participants in the Spirit's ministry of consolation and solidarity, extending divine comfort across the boundaries of species. Future research could further explore this pneumatological

Marius Nel adds another dimension that strengthens this vision. His account of the Paraclete underscores the functional unity between Jesus and the Spirit, portraying the Spirit as both advocate and guide, particularly in contexts of hostility and unbelief.[55] This aligns with Wallace's vision of the Spirit as a "comforter," "advocate," and "healer," but Nel's forensic framing adds an additional dimension: the Spirit does not merely comfort and advocate for creation passively but actively testifies on behalf of the voiceless, including non-human creatures, within the cosmic courtroom of divine justice. This is where, I suggest, pneumatology gains both edge and depth: the Spirit is not simply present in creation's suffering but also bears witness against the structures that produce it.

Wallace's view of the Spirit's solidarity with the marginalised, including animals, is closely related to Nel's portrayal of the Paraclete as one who ensures that disciples are not left "orphaned" (John 14:18) in their mission to witness to Christ amidst a world of systemic oppression and exploitation. Extending this idea to creation, the Spirit's advocacy for animals, as voiceless victims of ecological and industrial harm, reflects the continuation of Christ's redemptive work through the Spirit's active presence. Just as the Paraclete lends weight and credibility to the testimony of the disciples, so the Spirit lends divine weight to the basic dignity and significance of animal life, advocating for their liberation and inclusion in God's redemptive purposes. The Spirit, through its advocacy and healing, reveals the divine solidarity embedded in creation's suffering. By collapsing the functional roles of Christ and the Spirit, Nel reinforces Wallace's assertion that the Spirit embodies and continues Jesus' mission of restoration and healing. For Wallace, this includes a broadened eschatological vision where animals are not merely passive recipients of divine comfort but active participants in the Spirit's reconciling work, guided towards the ultimate renewal of all creation.

In this way, Nel and Wallace together present a pneumatology that is both forensic and pastoral: the Spirit convicts the world of its complicity in ecological harm while simultaneously advocating for and comforting creation's most vulnerable members. This dual emphasis provides a robust

reciprocity, examining how the Spirit's presence might facilitate a deeper, more integrated understanding of human-animal relationships within the context of divine justice and eschatological hope. See Charles, "'Animals Just Love You as You Are.'"; Walsh, "Human-Animal Bonds I."

[55] Nel, "The Notion of the Holy Spirit as Paraclete from a Pentecostal Perspective," 6.

framework for expanding the theological imagination to include animals within the Spirit's redemptive, reconciling, and healing work.

Read together, Moltmann, Wallace, and Nel offer more than parallel insights; they provide a theological grammar that allows us to speak of the Spirit as both forensic and pastoral, confronting ecological harm while standing in solidarity with creation's most vulnerable. This convergence invites a theological imagination capacious enough to hold together advocacy, judgement, and healing. In my view, such a pneumatology presses the Church towards concrete ethical and theological reorientation: if the Spirit is advocate and healer of all creation, then the Church cannot remain complicit in systems that silence animal suffering.

This vision also refuses to relegate eschatological hope to abstraction. It demands present-tense participation. By adopting ethical and legal reforms that reflect the values of the coming Kingdom, the Church can participate in the Spirit's work of liberation. As Moltmann reminds us, the new creation will not only bring the freedom of the children of God but will also encompass the deification of the cosmos, where all beings partake in the life and love of the Creator. This vision of eschatological hope demands a holistic theological response that integrates divine justice, Spirit-led advocacy, and the healing of all creation.

The Grieving Spirit: Extinction as a Theological Problem

The unprecedented rate of species extinction and the stark decline in global wildlife abundance testify to the devastating consequences of human action on the earth.[56] From a pneumatological perspective, this crisis is not only ecological but spiritual: the Spirit can be understood as grieving over the loss and suffering that permeate creation.[57] The Spirit

[56] Feminist theologian Rosemary Radford Ruether addressed the theological implications of species extinction in 1992.See Radford Ruether, *Gaia and God*, 100–102.

Current assessments of extinction rates, though varied, suggest a rate at least 100 times higher than the natural background extinction rate over the long term. Current data paints a concerning picture, revealing an average decline of 69% in species abundance between 1970 and 2018, with wild animals now constituting only a mere 4% of the total global biomass. Consequently, the ongoing ecological crisis has been labelled as the sixth mass extinction event. So, Almond, Almond, "Living Planet Report 2022: Building a Nature-Positive Society," 31.

[57] This section is partially taken from my forthcoming publication, Rizzo, "Beyond Eco-Mission."

is not merely the giver of life but also the compassionate presence who mourns the world's degradation.[58] In this light, extinction registers not only as a biological fact but as a theological wound—a rupture that reverberates through divine sorrow. The Spirit's grief over the destruction of ecosystems and the loss of species reveals the depth of divine investment in creation's flourishing. Such an interpretation reframes the ecological crisis as a moment of spiritual reckoning, summoning humanity to renewed responsibility in solidarity with the grieving Spirit.

In the Anthropocene era, where human actions significantly influence the planet's fate, crises manifest in various forms, from problematic circumstances to critical decision-making moments.[59] For Sigurd Bergmann, extinction (in the context of anthropogenic climatic change) can be regarded as human sin due to its connection to the broader human failing of fetishising power. Such a failing leads to the disenchantment of the sacredness inherent in earth and life, reducing animals to mere instruments for dominance.[60] From a Christian ecological perspective, recognising this distortion of power enables the discernment of the Holy

[58] Moltmann would argue that every experience of creation is inherently an experience of the Spirit itself. Therefore, when animals experience suffering, whether through factory farming, habitat destruction, or the threat of extinction, such suffering is felt within the Spirit. Moltmann, *The Spirit of Life*, 35. This perspective challenges the view of animal experiences as isolated events, instead presenting them as deeply intertwined with the Spirit's own experience and calling for a compassionate response that honours the Spirit's presence in all living beings.

[59] The term Anthropocene combines *anthropos* (human) and *kainos* (new, recent), suggesting a geological epoch marked by humanity's significant influence on Earth's geological layers. It signifies the transition from the Holocene, characterized by agricultural and societal development, due to profound human impacts on planetary processes. Ernst M. Conradie, *Secular Discourse on Sin in the Anthropocene: What's Wrong with the World?*, Environment and Society Series (Lanham: Lexington Books/Fortress Academic, 2020), 1. Craig Ritchie makes note of Indigenous and decolonial scholars who have critiqued the discourse surrounding the Anthropocene for its overarching "one-world" framework and fatalistic undertones. Indigenous scholars argue that the notion of a singular future "end of the world" and the idea that "we are all in it together" amplify colonial narratives that portray extinction or eradication as inevitable for indigenous peoples and other marginalized communities. Additionally, they contend that this discourse obscures the underlying social conflicts of the Anthropocene, including the reality that the world currently facing its demise was only made possible by the destruction of numerous other worlds, the repercussions of which continue to be felt today. See Ritchie, "Extinction in the Anthropocene," 137.

[60] Sigurd Bergmann, "Where on Earth Does the Spirit 'Take Place' Today? Considerations on Pneumatology in the Light of the Global Environmental Crisis," in *Christian*

Spirit's presence in the struggles of all living creatures, including those facing extinction.

Anita Davis deepens this insight by interpreting sin relationally. Sin, for her, is the disruption of the Creator–creation relationship—a disruption that opposes Pentecost's transformative vision of communion.[61] I find Davis's ecofeminist lens especially generative for pneumatology. By linking the oppression of women and non-human creatures, she exposes the structural denial of interconnectedness with God that undergirds ecological harm. This reading underscores that extinction is not merely the tragic end of species but a relational fracture within the community of creation. Pentecost, in contrast, names a vocational call to restored relationality through the Spirit. Davis states, "the relational relativity of Creator and creation in Christ by the Spirit and the vocational call to humanity through Pentecost provides the basis for understanding the current state of creation."[62] This theological contrast sharpens the stakes: understanding extinction through Davis's lens calls us to treat our non-human neighbours not as commodities but as fellow participants in God's life.

Matthew Eaton takes the discussion further by treating extinction as an eschatological tragedy. What I find striking in Eaton's work is his insistence that extinction reverberates within the divine life itself. By describing extinction as "ecocide," Eaton draws a bold theological parallel with deicide, interpreting ecological collapse as a disruption of the world's creative harmony.[63] His cruciform reading is particularly provocative: the cross becomes not merely a symbol of human redemption but a divine protest against violence—whether imperial or ecological. This reframing invites us to see extinction not simply as a natural process or moral failing, but as a site where divine grief, protest, and justice converge. In Eaton's view, Christ's embodiment emphasises the sanctity of physical existence, elevating creaturely concerns and experiences to the realm of the divine. A crucial aspect of this is the understanding of suffering, which plays a significant role in defining divine identity and experience. The pivotal

Faith and the Earth: Current Paths and Emerging Horizons in Ecotheology., ed. Ernst M. Conradie. (New York: Bloomsbury T&T Clark, 2014), 62.

61 Davis, "A Pentecostal Theological Perspective on Ecofeminism."

62 Davis, 62–63.

63 Eaton, "Ecocide as Deicide," 363.

event in shaping this understanding within Christian tradition is the crucifixion, where Jesus' death is interpreted as a rejection of the oppressive power and violence wielded by the Roman Empire. Jesus stands in opposition to the brutality used by the elite to maintain control over diverse communities, particularly the marginalised. Therefore, the cross is not seen as a passive acceptance of the fate of vulnerable bodies but rather as a divine protest against the sovereignty of the powerful at the expense of the weak. Jesus, embodying human vulnerability, becomes a champion for the oppressed, and his life and death follow a pattern of advocating for justice, especially when bodily integrity is violated. This concept, termed "cruciformity," suggests that God incorporates experiences of injustice, violence, and the subsequent ethical imperative for liberation into divine being. Cruciformity embodies a divine stance against violence, carrying forward into the divine essence in an eschatological sense. Eaton concludes with an ecological interpretation of Matthew 25, emphasising that human actions towards earth and its creatures are inseparable from their actions towards God.[64]

In the context of animals, pneumatology, and the concept of the grieving Spirit, extinction becomes not only a biological and ecological crisis but also a profound theological disruption. Extinction, as a manifestation of environmental degradation and ethical failure, represents a breakdown in the relationship between humanity, other creatures, and the Creator. This rupture speaks to a failure of stewardship, wherein human actions, rooted in exploitation and disregard for other life forms, have disrupted the harmony intended within creation.[65] From a pneumatological perspective, the Spirit is understood as grieving alongside creation,

[64] Eaton, 363.

[65] Christopher Southgate's evolutionary-theodicy framework recognizes pre-human extinctions as integral components of the earth's evolutionary history, contributing to the diversity of life. He views them as part of the dynamic and often unpredictable nature of evolution, shaping the trajectory of life on earth. Similarly, human-induced extinctions represent a continuation of this process, albeit with a significant difference: they are driven by human activity rather than natural forces. While pre-human extinctions occur as a result of environmental change and competition, human-induced extinctions stem from factors such as habitat destruction, pollution, and overexploitation of resources. Southgate contends that these activities disrupt ecosystems and accelerate extinction rates beyond what would occur naturally. Unlike pre-human extinctions, which are part of the inherent dynamics of evolution, human-induced extinctions reflect a deviation from ecological balance and pose a threat to the stability of ecosystems worldwide. In comparing pre-human and human-induced extinctions, Southgate underscores the ethical dimension

mourning the loss of biodiversity and the diminishment of life's diversity, which reflects God's creative work.

Tony Richie's insistence that divine healing ultimately points towards an eschatological redemption of all creation is a crucial starting point. His argument resonates strongly with Pentecostal pneumatology: if the Spirit's work is cosmic, then healing cannot remain anthropocentric. Frank Macchia sharpens this critique by exposing how individualistic conceptions of healing neglect systemic injustices, including the pervasive suffering embedded in ecological degradation. What is compelling here is Macchia's call for a pneumatological imagination capacious enough to hold together human and non-human liberation—a theological widening that I see as essential for a Pentecostal engagement with ecological and animal suffering.

The Wounded Spirit: Healing Creaturely Relationships

Pentecostalism, with its holistic emphasis on healing, has traditionally encompassed the spiritual, emotional, and social dimensions of human experience. Yet this theological vision can—and, I contend, must—be widened to embrace the non-human creation.[66] Tony Richie's insistence that divine healing ultimately points towards an eschatological redemption of all creation is a crucial starting point. His argument resonates strongly with Pentecostal pneumatology: if the Spirit's work is cosmic, then healing cannot remain anthropocentric.[67] Frank Macchia sharpens

of the latter. While both types of extinctions shape the evolutionary process, human-induced extinctions raise moral questions about humanity's stewardship of the earth and its responsibility to preserve biodiversity. From Southgate's perspective, addressing human-induced extinctions requires not only scientific understanding but also ethical reflection and concerted efforts to mitigate the impacts of human activity on the natural world. See Christopher Southgate, *The Groaning of Creation: God, Evolution and the Problem of Evil* (Louisville, KY: Westminster John Knox Press, 2008), 124–33. See also Elizabeth A Johnson, *Ask the Beasts: Darwin and the God of Love* (London: Bloomsbury Continuum, 2014), 248–53. Yong advances this perspective by contextualising Romans 5 within the paradigm of evolutionary geology to account for animal death before the emergence of humans. See Amos Yong, *Renewing Christian Theology: Systematics for a Global Christianity* (Waco, TX: Baylor University Press, 2014), 275–76.

[66] Swoboda echoes this point that Pentecostal understandings of restoration and reconciliation can be reimagined to address the ecological and non-human dimensions of God's redemptive work. Swoboda, *Tongues and Trees*, 246.

[67] Richie, "Radical and Responsible," 226.

this critique by exposing how individualistic conceptions of healing neglect systemic injustices, including the pervasive suffering embedded in ecological degradation.[68] What is compelling here is Macchia's call for a pneumatological imagination capacious enough to hold together human and non-human liberation—a theological widening that I see as essential for a Pentecostal engagement with ecological and animal suffering.

This widening finds further traction in A.J. Swoboda's engagement with Mark Wallace's ecotheology. Swoboda identifies in Wallace's pneumatological vision a framework that binds ecological ethics to the Spirit's sustaining work.[69] I see this as an opportunity for Pentecostal theology to shed any residual dualisms that separate the spiritual from the ecological. By envisioning the Spirit as both life-giver and reconciler of creation, Wallace provides Pentecostal thought with a conceptual grammar for extending its healing mission beyond the human. This is not simply about adding ecology to an existing framework; it is about recognising that the Spirit's reconciling activity fundamentally shapes how Pentecostals might imagine healing in a groaning creation.

Wallace's ecotheology rests on two key convictions: that the Spirit animates all life and that the sacredness of creation demands an ethical response of healing. Here, I find his emphasis on repentance particularly significant. By positioning repentance as ecological, Wallace invites Christian communities to participate in healing the fractured bonds between human and non-human creatures.[70]

This pneumatological vision disrupts the tendency to see the Spirit only in terms of flourishing; Wallace's inclusion of the Spirit's "dark side" — as judge and agent of upheaval within ecological crises—pushes Pentecostal theology to grapple with the Spirit's role amid environmental catastrophe.[71]

His notion of the "wounded Spirit" is especially generative for animal theology. By locating the Spirit's suffering within creation itself, Wallace collapses the distance between divine life and creaturely pain. This theological move creates space to speak of animals not merely as recipients

[68] Macchia, *Baptized in the Spirit: A Global Pentecostal Theology*, 86.

[69] Swoboda, *Tongues and Trees*, 135.

[70] Swoboda, 135–36.

[71] Swoboda, 136. The Spirit is active not only in life's flourishing but also in the upheaval of ecological crises, symbolised by storms, earthquakes, and other natural processes.

of divine action but as participants in a shared economy of suffering and hope. Here, the suffering of animals and the broader creation becomes entangled with the suffering of God's Spirit, deepening theological reflection on divine vulnerability. Wallace's imagery of the Spirit embodied in natural organisms and processes expands pneumatological imagination to include the Spirit's enfleshment within creation's pain and joy. This pneumatological vision, while not explicitly Pentecostal, is closely aligned with Pentecostal theology's emphasis on the Spirit's dynamic presence within the economic Trinity.[72] Swoboda's retrieval of Wallace's insights shows Pentecostal theology's potential to become a robust ecological ethic grounded in the Spirit's dynamic and healing presence.

This is where practical theology intersects with pneumatology. If the Spirit is healer and reconciler, then Christian praxis must reflect that vocation towards all creation. Practical theology becomes the site where immediate ecological crises meet eschatological vision. It demands both reactive measures to address pressing threats like climate change and proactive habits that anticipate a healed creation. This trajectory can be used to develop an animal theology within a Pentecostal perspective. Recognising animals as valuable to God invites the global Church into deeper stewardship and compassion. Such a vision not only enriches contemporary pneumatology but also fosters sustainable practices that honour all living creatures, aligning theological commitments with tangible care for creation.

However, addressing the broader ecological crisis necessitates moving beyond the Spirit's healing role to interrogate the ethical responsibilities of humanity towards non-human creation. While Christian environmental ethics often addresses extinctions as indicative of larger issues such as industrialism, modernity, or colonialism, specific attention to non-human lives and deaths remains minimal within Christian discourse.[73] Andrew Linzey echoes this sentiment and affirms that "Christians entrusted with a ministry of reconciliation to the whole of creation need to be credible signs of the Gospel for which all creatures long."[74] This ministry of reconciliation invites a shift in practical theology to encompass not only human-focused concerns but also the deeply intertwined realities of

[72] Swoboda, 141.

[73] Jenkins, "Loving Swarms: Religious Ethics amid Mass Extinction," 20.

[74] Linzey, *Animal Gospel*, 15.

ecological degradation and non-human suffering. Practical theology operates at the intersection of crisis and opportunity, navigating the delicate balance between present realities and potential futures. Practical theologians are tasked with delving deeper into these dynamics of change, addressing ingrained habits that contribute to ecological degradation alongside traditional human-focused crises. This expanded approach calls for urgent reactive responses to immediate threats like the global climate crisis, while also initiating proactive and anticipatory actions to safeguard the planet's future.

The eschatological vision of creation's redemption further enriches this account, offering models of life that embody the gospel for both human and non-human creatures. In discussing the eschatological implications of living in harmonious relationship with creation, Tripp York draws upon the example of Saint Francis of Assisi to illustrate a model of life that embodies the gospel not just for humans, but for all creatures. York notes that Francis' life served as "good news" for all creation as he sought to live in a way that reflected Christ's love and care for all beings, ultimately seeing himself in covenantal partnership and kinship with them.[75] Francis' intentional kindness, avoiding harm to creatures like worms and feeding insects in winter, was not motivated by mere sentimentality; rather, it stemmed from his understanding of creation's divine purpose and inherent goodness. Francis viewed humanity's role as a unique calling to name and care for animals in a manner that mirrored Jesus' care for humanity, resisting a utilitarian view that reduces animals to objects for human use.

York suggests that an eschatological approach to life calls Christians to reconsider how they name and treat animals, advocating for a language and perspective that sees animals as covenant partners rather than resources to be exploited. This eschatological "naming" acknowledges animals in light of their place in God's creation and the ultimate purpose for which they were made. Living eschatologically is intrinsically linked to healing broken relationships between humans and non-human creatures. Extending York's vision, I propose that a Spirit-filled life

[75] York, "Can the Wolf Lie Down with the Lamb without Killing It? Confronting the Not-So-Practical Politics of the Peaceable Kingdom," 158. York reiterates this idea in his book, Tripp York and Laura Hobgood-Oster, *The End of Captivity?: A Primate's Reflections on Zoos, Conservation, and Christian Ethics* (Eugene, OR: Cascade Books, 2015), 112.

empowers believers to participate in this creaturely reconciliation now. Pentecostal pneumatology, with its emphasis on the Spirit's dynamic, transformative presence, provides precisely the theological energy needed for this task.

In this way, Pentecostal theology can move beyond simply affirming the Spirit's role in human healing to articulating a comprehensive pneumatology of creation—one that positions the Spirit as life-giver, reconciler, and sustainer of all creatures. This constructive vision integrates healing, repentance, and eschatological hope, offering both a theological and practical framework for engaging ecological degradation and affirming animal life within God's redemptive purposes.

Conclusion

Eschatology, when seen through a pneumatological lens, stretches far beyond human horizons. It gathers animals into the wide arc of divine redemption, refusing to leave any corner of creation untouched by the Spirit's reconciling breath. Drawing on Moltmann's vision of eschatological justice, I have traced how the Spirit's life-giving, healing, and advocating presence unsettles anthropocentric habits of thought and opens up a more expansive theological landscape. Within this vision, redemption is neither narrowly spiritual nor exclusively human; it is cosmic, creaturely, and intimate.

This broadening requires a reckoning with how eschatology has often been framed. Dualistic patterns that sever the present from the future risk treating creation as disposable—a mere prologue to some spiritual finale. Yet the Spirit's work invites a subtler balance. Continuity affirms the goodness and worth of the material world; discontinuity acknowledges the need for transformation in the face of sin, violence, and decay. The Kingdom is both already here and still coming, holding these tensions in creative suspense.

Pentecostal theology, with its dynamic vision of the Spirit's presence, is well placed to inhabit this tension. Its eschatology can stretch to include animals, not as afterthoughts, but as fellow participants in the unfolding story of renewal. This is more than a theological adjustment; it is a shift in posture—a way of seeing creation as the Spirit does: alive, wounded, waiting, and destined for glory.

Such a vision doesn't remain on the page. It presses towards ethical response, inviting communities to live now in light of the world to come.

The Spirit's advocacy for all creation summons the Church to care, to listen, and to act in ways that anticipate divine justice. Animals, in this telling, are not merely recipients of redemption; they stand within it, bearing witness to the Spirit's ongoing work.

And it is here, in this mingling of present groaning and future glory, that we turn to the voices of creatures themselves. The next chapter explores animal glossolalia—not as a curious aside, but as a glimpse of the harmony to come, even as it reverberates through the fractured present.

CHAPTER 6

Animal Glossolalia

Having explored the Spirit's work in creation, covenant, soteriology, and eschatology, this chapter turns to another manifestation of the Spirit: glossolalia. The preceding discussions have traced the Spirit's pervasive presence across all creation—from the breath of life (*ruach*) that animates all creatures, to the Spirit's redemptive embrace in soteriology, to the eschatological hope of a reconciled creation. These explorations have affirmed that the Spirit's activity is not limited to human beings but extends to all flesh, encompassing non-human creatures as integral participants in the divine economy of creation and redemption.[1]

Building on this theological foundation, I now consider glossolalia as a sign of the Spirit's indwelling presence and its implications for the broader creation. Traditionally understood within Pentecostal theology as the Spirit-empowered gift of tongues, glossolalia is seen as a human expression of worship and communication with God. However, if the Spirit's baptism truly extends to "all flesh," as argued in this book, glossolalia must also be reframed through a wider lens. Might the vocalisations, instinctual responses, and communicative behaviours of animals represent Spirit-infused expressions of praise, lament, or intercession? This chapter

[1] A version of this chapter has been published, Rizzo, Daniela. "Animal Glossolalia: A Pneumatological Framework for Animal Theology." *Pneuma* 46, no. 1 (2024): 60–79. https://doi.org/10.1163/15700747-bja10108.

D. Rizzo, *Animal Pneumatology*, The Palgrave Macmillan Animal Ethics Series, https://doi.org/10.1007/978-3-032-18723-9_6

explores the possibility that animals, as Spirit-baptised participants in creation, contribute their unique voices to the cosmic chorus of worship. The chapter aims to broaden the scope of Pentecostal pneumatology and to deepen our understanding of the Spirit's work in all creation.[2]

While I recognise the broader ecotheological approach that includes the voices of all creation (such as forests, rivers, etc.), my focus narrows specifically to the sentience of animals. My aim is to explore, through pneumatological imagination, how animals may express their voice in ways that move beyond their natural behaviours. I am particularly interested in how animal voice might be interceded by the Spirit, possibly through what I am calling animal glossolalia. This envisions animals, as sentient beings, participating in a deeper spiritual communication, where their experiences and voices are not only heard but also transformed or amplified by the Spirit. Seen this way, the Spirit might enable animals to communicate in ways that transcend their natural capacities.

In this chapter, I propose a framework that envisions non-human animals as participants in glossolalic praise and lament, so that their voices are woven into the broader fabric of cosmic worship. I argue that the sighs

[2] Ecotheologians have widely explored the notion of "voice" in the context of ecological and theological discourse, particularly through initiatives like the Earth Bible Project. This scholarly endeavour, spearheaded by the Earth Bible series and edited by Norman Habel, approaches the biblical text from an ecological perspective. One of the key features of the Earth Bible Project's approach, as it has developed over the past 20 years, is the creative retrieval of the Earth's voice (see Norman C. Habel, *Readings from the Perspective of Earth* (Sheffield: Sheffield Academic Press, 2000), 24; 46–48.). This method focuses on reclaiming and amplifying the voice of the Earth, similar to the creative articulation phase seen in earlier feminist biblical interpretation.

In this project, the concept of "voice" transcends human speech and extends to the expressions, actions, and presence of all living entities within creation. These non-human voices, including those of animals, plants, and even the Earth itself, are understood to have moral and theological significance, calling for a revaluation of how creation is perceived and valued. In this project, the concept of "voice" transcends human speech and extends to the expressions, actions, and presence of all living entities within creation.

Balabanski's paper engages with this broader ecotheological framework by revisiting the Earth Bible Project's principles, highlighting that all beings—both human and non-human—are interconnected within the ecological order, Vicky Balabanski, "Revisiting the 'Voice of Earth.' Earth Bible Principle #3," *Journal of Ethics in Antiquity and Christianity* 5 (2023): 75–86, https://doi.org/10.25784/jeac.vSil.1204. This interconnectedness implies that non-human animals have their own "voice" within the ecological system. This voice does not necessarily manifest in language or verbal communication but in the ways animals exist, interact, and experience the world, particularly as sentient beings capable of suffering and experiencing joy.

and groans of animals, whether comprehensible to humans or not, serve as prayers and praises, interceded by the life-giving Spirit. These utterances, carried on the breath of creation, bind animals with humans as eschatological partners. Expanding this vision, I introduce the notion of animal glossolalia, where non-human creatures are Spirit-baptised participants in the shared eschaton, bearing witness to and actively participating in divine communion.

Through their sighs and groans of lament, animals engage in a kind of prayer to the Creator. This exploration draws on Macchia's work on glossolalia, positioning it as a key conversation partner, and seeks to broaden the pneumatological understanding of glossolalia to include species beyond humanity. If glossolalia transcends human boundaries of race and culture, as an unintelligible yet Spirit-filled form of prayer, might it also cross species lines? In this light, glossolalia becomes a universal means of divine connection, accessible to all creatures who partake in the breath of life.

Sentience and Animal Spirituality

Building on ideas presented in the section on animal ontology, it is important to highlight key aspects from prior discussions on animal sentience and spirituality as they lay the foundation for advancing my argument in this chapter. I[3] have already established that animal sentience involves the capacity to experience both positive and negative emotions and exhibit behaviours reflective of inner life. Recognising that animals are subjects with autonomy and integrity rather than mere mechanisms underscores their role as companions in creation, reconciliation, and redemption. This view connects directly to pneumatology, as the Spirit animates life and unites all creatures, human and non-human, in fellowship with God. Pannenberg's pneumatology supports this by affirming that the Spirit's presence extends throughout all creation, giving life, movement, and activity to non-human creatures.[4] This connection between the Spirit and animals emphasises their participation in a divine framework, where sentient beings, through their existence, engage in creaturely praise and lament, as evidenced by various biblical texts.

[3] See Chapter 2.

[4] Pannenberg, *Systematic Theology Vol 2*, 32.

In light of these earlier arguments, my focus now turns to how this sentience and spiritual relationship can be understood through a pneumatological imagination. This exploration parallels, in some ways, the reflective natural encounters in works like *Pilgrim at Tinker Creek,* where observation of the natural world leads to insights about life, existence, and God.[5] However, unlike the anthropocentric perspective often adopted in nature writing, I want to amplify the spiritual dimensions of non-human voices, recognising them as vital expressions within the broader, Spirit-filled creation.

Glossolalia: Beyond Human Experience

Animal glossolalia, as a pneumatological idea, extends Swoboda's metaphor of eco-glossolalia and weaves this metaphor into the very fabric of creation itself. Animals, far from being passive bystanders, are active participants and witnesses in the shared eschaton, their sighs and groans borne on the breath of the life-giving Spirit, rising like a prayer to the Creator. By marking animals as sentient beings, linking them intimately to their human counterparts, I sketch with a broad stroke, one that includes all living beings as sentient co-creatures. Animal participation in praise and prayer under the umbrella term of glossolalia can be pneumatologically woven into existing glossolalic definitions of Macchia and Swoboda, to assist in the construction of a species-inclusive glossolalia.

Swoboda's term *Eco-Glossolalia* metaphorically refers to the Spirit-baptised creation as a baptism of "all flesh," with non-human creatures as eschatological partners.[6] Building on Swoboda's ideas, I propose that the Spirit's work within creation can extend to animals, enabling them to participate in the divine economy through groans and sighs that may serve as forms of prayer or praise. However, I do take into consideration that not all animal sounds can be regarded as glossolalia.[7] This raises the critical question of discerning the Spirit's activity, a principle equally

[5] Anne Dillard, *Pilgrim at Tinker Creek* (New York: Harper Perennial Modern Classics, 2013).

[6] Swoboda, *Tongues and Trees*, 113.

[7] This is not to say that when a cat meows that the mammal is necessarily engaging in prayer, as they may simply be vocalising a desire to be fed, or to be let outside. Although, for the psalmist in Psalm 104:21 the lion's incessant roar during the night is interpreted as a plea to God for sustenance. As Eaton explains, the psalmist envisioned a world where

applicable to human forms of glossolalia. Moltmann's concept of a nature hermeneutic is helpful here. It suggests that nature speaks to humanity through a "language of signs."[8] This hermeneutic implies that animal vocalisations, while not always prayerful, can be viewed as communicative acts that reflect their deep connection to the Spirit and creation's longing for renewal (Rom 8:22–23).[9] For Moltmann, deciphering this "language of signs" involves attentiveness to the Spirit's activity in creation, much like interpreting an unknown language. This allows for the possibility that animal expressions, while primarily functional, may also carry layers of spiritual significance when viewed through a pneumatological lens. Thus, discerning the Spirit's work in animals requires an interpretive framework that respects their creaturely nature while remaining open to the Spirit's transformative presence within them.

I begin by positioning glossolalia as a cosmic Spirit-language and reimagine this as a pneumatological experience that extends beyond the human experience. The act of worship finds expression across all forms of creation, serving as a means to usher in the Kingdom of God.[10] The concept of cosmic worship, embodied as a form of glossolalia, becomes a manifestation of the Spirit's role in bringing forth the Kingdom of Heaven on Earth. A celestial language, accessible to all creatures, becomes a unifying element in this divine process. While human beings, bearing the *imago Dei*, possess the autonomy to decide whether or not to engage with their Creator and to pursue the gift of tongues through the Holy Spirit, non-human creatures are intrinsically bound to the cosmic community of Spirit engagement. Their creaturely nature entwines them with the broader, Spirit-breathed creation, making them natural participants in this cosmic glossolalic chorus. This creaturely-shared language of the Spirit

all creatures, including the mighty lion, relied on God for their survival. John Eaton,. *The Circle of Creation*, 52.

[8] Jürgen Moltmann, *Sun of Righteousness, Arise!: God's Future for Humanity and the Earth* (La Vergne, TN: Hymns Ancient & Modern Ltd, 2010), 126.

[9] As Paul writes in Rom 8:22, all creation, including humans, groan *inwardly*, and it is the Spirit that intercedes on all of creation's behalf.

[10] Isaiah's vision of the peaceable kingdom reinforces the idea that triumph is a collective endeavour, involving every creature.

becomes a bridge, linking all of creation in the continuous flow of divine communication.[11]

Kenneth Archer defines glossolalic prayer as "Spirit-inspired praise and lament."[12] David Perry adds that "the relationship between Spirit baptism and tongues may be understood theologically as involving the communication of elemental meaning and expressing the inexpressible."[13] Mark Cartledge similarly describes glossolalia as a yearning for the consummation of God's Kingdom on earth.[14] Frank Macchia frames it as a form of "unclassifiable, free speech in response to an unclassifiable, free God."[15] He further emphasises that glossolalia must be understood eschatologically, where the unintelligible cries represent "yearnings for the redemption and liberation to come," encompassing all creation.[16]

[11] In Chapter 2, I argued that animals, by virtue of their ontology, possess an inherent and unbroken connection with their Creator. Consequently, free will is not required of animals, as their very existence—being what they were created to be—constitutes an act of worship to God. I do not apply the concept of free will to animals in the same way it is attributed to humans created in the *imago Dei*. Instead, animals, by their nature, are oriented toward their Creator and fulfil their purpose simply by existing as they were designed, without the need for the exercise of free will as a defining characteristic of their relationship with God.

[12] Kenneth J. Archer, *God - Creation's Hope, Creation - God's Home*, 209. On the idea of glossolalia as lament, Stephen C. Torr states that "it is possible to suggest in the light of Rom. 8.26 that it is an example of a heartfelt cry – a groan perhaps! – out of the depths and the felt emotion of the situation, to God, that fractures the boundaries of identifiable human language." Stephen C. Torr, "Lamenting in Tongues: Glossolalia as a Pneumatic Aid to Lament," *Journal of Pentecostal Theology,* 26(1), 30–47, 2017, 43, https://doi.org/10.1163/17455251-02601005.

[13] David Perry, *Spirit Baptism: The Pentecostal Experience in Theological Focus*, Global Pentecostal and Charismatic Studies (Boston: Brill, 2017), 233, https://search.ebscohost.com/login.aspx?direct=true&db=nlebk&AN=1565681&site=ehost-live.

[14] Cartledge, *Charismatic Glossolalia: An Empirical-Theology Study*, 188.

[15] See Frank D. Macchia, "Sighs Too Deep for Words: Toward a Theology of Glossolalia." *Journal of Pentecostal Theology* 1, no. 1 (1992): 47–73. According to Macchia, glossolalia is evidence of the believer's participation in the divine life and an expression of the Trinitarian nature of God.

[16] Macchia, Sighs, 61–62. Yong, in his interaction with Macchia's theology of glossolalia, agrees with this definition and further suggests that a community of believers who experience glossolalia shares in "divine freedom, divine speech, and divine life. Amos Yong, "The Truth of Tongues Speech: A Rejoinder to Frank Macchia." *Journal of Pentecostal Theology* 6, no. 13 (October 1998): 114. Also, see 189. For Cartledge, glossolalia occurs when both God and humanity are speaking simultaneously, with the human uttering unintelligible words while God provides the content of the speech. The act of glossolalia is

While Macchia argues that glossolalia is the explicit language of the *imago Dei*, I propose that animals can also participate in this divine dialogue.[17] Through the Spirit, animals communicate their praise and lament as the Spirit intercedes on their behalf. Unlike humans, who interpret and discern the transformative power of glossolalic symbols, animals express their own instinctive "sighs and groans"—prayers of praise, lament, and yearning for liberation—through the Spirit, who carries their voices to the Creator.[18]

Tongues of Angels and the Cosmic Chorus

Glossolalia is referred to as a language, or "tongues of angels" (Acts 2:1–4; 1 Cor 12:10, 13:1). It is conceived as a cosmic language of the Spirit, where the "tongues of angels" is a unique creaturely tongue that harmoniously contributes to the symphony of cosmic worship within creation.

Richard Bauckham's scholarship on the four hybrid angels is instructive in this connection. John's visionary encounters in Revelation 4:6–9 and Revelation 5:6–14 offer glimpses of celestial beings encircling the divine throne in heaven. These creatures, described with distinct features (a lion, an ox, a human face, and an eagle), may, according to Bauckham, symbolically represent various categories of animate creation, mirroring the four earthly creatures. I propose that these angelic beings partake in what can be termed as "tongues of angels," a form of cosmic glossolalia that resonates with expressions from both human and non-human entities. This posits a collective, transcendent language transcending conventional verbal communication, prompting a broader perspective on this phenomenon. Bauckham underscores that non-human creatures, despite their divergence from human consciousness, should be regarded as active

seen as an extension of the indwelling presence of the Holy Spirit, representing symbolic participation in "an abiding communion with God." Just as glossolalia is described as a language of divine mystery where God and humanity speak in the same utterance, a similar concept can be extended to animals.

[17] Macchia, *Sighs Too Deep for Words*, 61; Macchia contends that "glossolalia can be seen as revealing something very profound about the human existence" before God. That everyone who is born of the Spirit is set free by the Spirit "in protest against any attempt" to oppress humanity from our freedom-giving God.

[18] Frank D. Macchia, "Discerning the Truth of Tongues Speech: A Response to Amos Yong." *Journal of Pentecostal Theology* 6, no. 12 (April 1998): 68.

participants in the act of worship.[19] His insights emphasise that many elements of creation, lacking the linguistic capacity for verbal expression or conscious intentionality towards their Creator, inherently contribute to magnifying God simply by fulfilling their intended roles. This fundamental contribution, stemming from their alignment with their Creator's design, inherently brings glory to God. Such creatures are, in essence, fellow members of the community of creation, sharing a common desire to worship their Creator.

Although the theme of creation's worship is not exclusively confined to a single scriptural passage, Bauckham suggests that Psalm 148 encapsulates this concept in a comprehensive way. Through a harmonious interplay of celestial beings and diverse categories of creatures, the psalmist beckons the entirety of creation to unite in an anthem of praise directed towards the Lord. The psalm paints a vivid tableau of creation itself, transforming into an orchestra where every element contributes its unique note to the symphony of God's praises.[20] This symphony, in turn, extends an open invitation to human participants.

A significant portion of worship depicted in the book of Revelation doesn't originate solely from Christians or even humans. Within the celestial realm, God receives adoration from various entities. The four living beings partake in this worship (Rev. 4:8; 5:8–9, 14; 19:4), as do angels (Rev. 7:11) and the entirety of heaven (Rev. 5:11; 7:11). A pervasive theme emerges as all of creation joins in worship (Rev. 5:13). This imagery serves to highlight the cosmic act of worship to the Creator. According to Bauckham's interpretation of the book of Revelation, the depiction of each living creature resembles each one of the four earthly creatures and may symbolise distinct categories of animate creation.[21] For instance, the creature resembling a lion could represent wild land animals, the one resembling an ox could represent domestic land animals, the one resembling an eagle could represent birds, and the one with a human-like

[19] Richard Bauckham, *Living with Other Creatures: Green Exegesis and Theology* (Waco, TX: Baylor University Press, 2011), 221–22.

[20] I also see this idea of the cosmic chorus echoed in Moltmann's work. See Jürgen Moltmann, *The Way of Jesus Christ: Christology in Messianic Dimensions*, trans. Margaret Kohl (Minneapolis, MN: Fortress Press, 1993), 289.

[21] John's vision in Rev 4:6–9 and Rev 5:6–14 includes the depiction of four living creatures surrounding the throne of God in heaven. These creatures are described with distinct appearances resembling a lion, an ox, a human face, and an eagle.

face could represent humans.[22] Interestingly, while one of the living creatures in Revelation has a human face, the other three are affiliated with other major categories of animal creation. For Bauckham, this portrayal suggests that humans do not hold a privileged or superior position among creatures.[23] The human-like creature is simply one among the four, not listed first or given precedence. As creatures participating in the worship of God, humans are not set above other creatures, but rather stand alongside them, sharing in the common worship of their mutual Creator.[24] This correlation suggests that each entity, regardless of its position or nature, plays a role in the broader cosmic worship of God.[25] The idea that humans stand alongside other creatures in worship, rather than being alone or elevated above them, reinforces the sense of unity and shared devotion in the cosmic worship paradigm.

To situate glossolalia in the context of cosmic language, I suggest that glossolalia functions as a means of communication that surpasses the limits of conventional linguistic confines. This concept underscores the idea that worship transcends the boundaries of human verbal expression. Just as glossolalia bridges the divide between the verbal and non-verbal realms, cosmic worship can be envisioned as a harmonious symphony of

[22] Richard Bauckham, "Creation's Praise of God in the Book of Revelation," *Biblical Theology Bulletin* 38, no. 2 (May 1, 2008): 60, https://doi.org/10.1177/01461079080380020201.

[23] Bauckham, "Creation's Praise of God,"62.

[24] Yong compliments this idea of a shared spiritual and material realm. He writes, "The spiritual realms are intertwined both ontologically and epistemologically. Regarding the former. The Spirit both hovers over the waters of creation and gives the breath of life; the human is intimately and intricately connected with the orders of creation." Amos Yong, *The Spirit Poured Out on All Flesh: Pentecostalism and the Possibility of Global Theology* (Grand Rapids, MI: Baker Academic, 2005), 300.

[25] Bauckham notes that the Christian interpretation of Rev 4:8 during the patristic and medieval periods revolved around Irenaeus' perspective, which posited that the four living creatures symbolise the four evangelists. These creatures were seen as representative of the universal nature of the fourfold gospel, aligning with the divisions of the earth or the four winds. However, an alternative viewpoint emerged within the Coptic and Ethiopian churches, where the traditional belief held that the living creatures served as heavenly representatives of earthly counterparts. This perspective, coexisting with Irenaeus' view, gained traction and is commemorated annually on November 4th within the liturgical calendar. This interpretation emphasises the living creatures' role as agents of petitionary prayer, rather than direct worship, while also highlighting their representative function, aligned with the four categories of earthly creatures. See Bauckham, *Living with Other Creatures*, 180.

expressions emanating from every facet of creation. This harmony finds resonance in the angelic beings partaking in the "tongues of angels," implying that even non-human creatures play a role in the grand chorus of worship.

John Poirier's research sheds light on an interesting connection between the liturgical *jubilus* and the concept of esoteric angelic languages. Building on this, I suggest that glossolalia could be understood as a form of hidden angelic communication, reflecting a deeper, clandestine dialogue within the divine realm.[26] This proposal gains further credence from the historical association of *alleluia*, a liturgical expression of praise in Christian worship, with angelic adoration.[27] This novel concept, termed *angeloglossy,* expands the existing frameworks of human glossolalic language and angelic esoteric languages. Poirier's work indicates how esoteric angelic languages could enrich our understanding of glossolalia, proposing that this Spirit-infused voice extends beyond human experience to encompass both celestial and terrestrial sentient beings.

Furthermore, the parallels between human and angelic languages, along with the introduction of the *angeloglossy* concept, suggest a more inclusive perspective on communication between diverse orders of creation and the divine. The language of worship transcends human boundaries, encompassing the full spectrum of sentient beings. Just as angels engage in worship through their distinct language reflective of their nature, so do animals, plants, and other forms of life express their devotion through their inherent characteristics.

There is a striking parallel in Paul's thought, where the focus shifts to the collective engagement of all creatures as depicted in Romans 8:20–22.

[26] Proirier's term "liturgical *jubilus*" refers to a joyful and exultant expression of praise and worship within the context of religious liturgy, particularly in Christian traditions. In the ancient world, there was a widespread belief that gods, angels, demons, and other divine beings spoke languages that were not used by humans, except in magical practices or ecstatic states. John C. Poirier, *The Tongues of Angels: The Concept of Angelic Languages in Classical Jewish and Christian Texts*, Wissenschaftliche Untersuchungen Zum Neuen Testament 2. Reihe (Tübingen: Mohr Siebeck GmbH & Co. KG, 2010), 1. In his book, Poirier highlights two key points. Firstly, Jewish and Christian sources exhibit varying views on the language spoken by angels, debating whether angels used Hebrew or had their own distinct language. Additionally, Paul's words in 1 Corinthians 13 and 2 Corinthians 12 cannot be attributed to any known Jewish or Christian sources, suggesting unique perspectives on angelic languages. Furthermore, Paul's writings generally have limited content on angels or demons.

[27] Poirier, *The Tongues of Angels*, 140.

For Paul, there exists an interdependence between nature and humanity, rooted in their shared identity as creatures of God, rather than a position of humans dominating nature. In a feminist reading of Romans, McGinn notes Paul's emphasis on eschatological vision and the obstacle preventing its realisation. The obstacle is not inherent to creation itself, but rather lies within the deficiencies of humanity. Human salvation remains incomplete, "and creation waits for its sibling."[28] Upon humans being embraced as God's children and inheriting God's freedom and glory, creation will similarly find fulfilment through the liberty and magnificence of God. Paul's insight into the ultimate purpose of creation is crucial; without it, his claims regarding creation's active role in human salvation would lack coherence.

Sighs to Deep for Words

In chapter 2 of this book, animals were described as capable of offering both praise and lament, which entails that they live in relationship with God and respond to divine love according to their own nature.[29] The implication here is that God hears the voices of non-human creation and, although animals do not possess reflective consciousness like humans, they still bear witness to the Creator's power. The establishment of God's eternal covenant with Noah and every living creature affirms that humans are not the only creatures that are addressed by God and called to live lives in response to God. In groaning solidarity with all creation, the same Spirit who dwells in the risen Christ intercedes on behalf of the entire created order, guaranteeing that, though creation is subject to decay, it will not be destroyed.

Though glossolalia is typically understood as a human experience, I will explore the possibility of a glossolalic faculty in Spirit-baptised animals as sentient beings. My model of glossolalia builds on Pentecostal theological emphases on Spirit-baptism and glossolalia, with a focus on understanding how this form of communication can be experienced across different species through the Spirit.

[28] Sheila E. McGinn, "All Creation Groans in Labour: Paul's Theology of Creation in Romans 8:18–23," in *Earth, Wind and Fire: Biblical and Theological Perspectives on Creation*, ed. Carol J Dempsey and Mary Margaret Pazdan (Collegeville, MN: Liturgical Press, 2004), 122.

[29] See Chapter 2.

By framing glossolalia as creaturely sighs and groans, I aim to find common ground in the glossolalic experience that can be applied across species. While humans can discern and interpret the transformative power of glossolalic symbols, the Spirit provides a spiritual voice by interceding through the instinctive "sighs and groans" of animal praise, lament, and yearning for freedom to the Creator. The concept of Spirit baptism in this way extends beyond a purely human experience of glossolalia and explores how the Spirit communicates and mediates with her non-human creatures. I will now explore these ideas in turn.

The task of revisioning the scope of glossolalia must begin with Frank Macchia's definition of glossolalia as vocal expressions akin to sighs and groans for the non-verbal dimension of glossolalic speech. In his article "Sighs Too Deep for Words: Toward a Theology of Glossolalia," Macchia refers to the biblical concept of "sighs and groans" to describe the experience of glossolalia, which he defines as a form of prayer that goes beyond the borders of human language. Macchia argues that glossolalia is a manifestation of the Holy Spirit, who intercedes with "sighs too deep for words" (Rom 8:26). He suggests that glossolalia allows believers to participate in the Spirit's intercession, as they offer prayers in a language that surpasses their own understanding, a language that can be expressed as sighs and groans.[30] Macchia highlights the significance of recognising glossolalia as a spiritual and transcendent occurrence rather than limiting it to a solely linguistic or psychological encounter.[31] He underscores the idea that our glossolalic prayers should harmoniously blend

[30] Macchia, *Sighs*: 59. One intriguing aspect of how early Pentecostals interpreted Rom 8:26 is the phenomenological descriptions provided by those who claimed to have personally experienced what Paul described. For instance, some interpreted Paul's use of the language of groaning in Rom 8:23, 24, and 26 as a literal expression of groaning that they themselves experienced. They believed that this groaning was most evident during times of deep prayer and intercession. In his article, "Inward Groans and Unknown Tongues," Stone reflects on an early account which explains how, "Much time was spent in intercessory prayer, and many groans and cries were heard as the Spirit prayed and groaned within, Rom. 8.26–27, according to the will of God." In such accounts, the groaning was understood as an audible experience, vocalised in their bodies through the agency of the Holy Spirit. See Jesse D. Stone, "Inward Groans and Unknown Tongues: Interpretations of Romans 8.26 in Early Pentecostal Literature," *Journal of Pentecostal Theology* 30, no. 1 (May 5, 2021): 90, https://doi.org/10.1163/17455251-30010007. In her book *God, Sexuality, and the Self*, Sarah Coakley integrates systematic theology with the subjective experience of contemplation. Over the course of her project, she addresses the similarities between silent prayer and tongues several times. Both involve "ceding to the Spirit" whereby the one who prays allows God to be the active subject by which the prayer operates. This is typified by the fact that both prayer styles involve a surrender of standard communication patterns in favour of either silence or tongues. Sarah Coakley, ed., "The Charismatic Constituency: Embarrassment or Riches?," in *God, Sexuality, and the Self: An Essay "On the Trinity"* (Cambridge: Cambridge University Press, 2013), 117, https://doi.org/10.1017/CBO9781139048958.006.

McGraw found that glossolalia was associated with decreased blood flow (perfusion) in the brain's frontal lobes. This decrease in blood flow aligned with the subjects' reported lack of intentional control over their glossolalia, suggesting that it was not a consciously controlled activity. The implications of this finding extend to theories about the frontal lobes' role in consciousness and volition, indicating that the reduced frontal lobe activity during glossolalia may support the notion that it is not consciously controlled. This finding may shed light on the neural mechanisms underlying altered states of consciousness and spontaneous vocalisations in religious or spiritual contexts. John J. McGraw, "Tongues of Men and Angels: Assessing the Neural Correlates of Glossolalia," in *Religion and the Body*, ed. David Cave and Rebecca Sachs Norris, Modern Science and the Construction of Religious Meaning (Brill, 2011), 61, https://www.jstor.org/stable/10.1163/j.ctv2gjx03h.7. Therefore, it could be argued that animals may experience transcendence through glossolalia, expressing their creaturely nature without consciousness being a necessary prerequisite for their participation. Even for those who remain unconvinced by current scientific and behavioural evidence of animal consciousness, this argument holds that their involvement in such divine expression does not hinge on the presence of conscious awareness.

[31] Within this context, Macchia describes Cyril Williams's view of tongues (or glossolalia) as a form of "mysticism of sound" that can represent a connection with the divine. Macchia, "Sighs," 52.

with the collective voice of the groaning creation. For him, glossolalia isn't detached from the world's suffering but rather serves as a heartfelt expression of yearning for the deliverance of a suffering creation.[32] When human freedom is experienced through the Spirit, it should be a freedom dedicated to the emancipation of all of creation.[33]

In situating eschatology as the context for Spirit-baptism, one could argue that unity with our co-creatures becomes a central aspect of God's divine renewal in the eschaton.[34] If glossolalic sighs and groans don't convey intellectual messages but instead stir the Spirit's transformative work within, then why should this profound experience be confined to only one species? Since God's covenant extends to every living creature, not exclusively to humans, it's plausible that all creatures engage in prayer and praise with the Creator Spirit through their unique vocalisations, including their sighs and groans.[35]

Animals, much like humans, possess distinct modes of communication that may elude human understanding. In a manner akin to the way humans communicate with God through glossolalia, one can envision the communication of animals taking on a similar participatory role. The concept of "primal spirituality," as defined by Harvey Cox, offers insight into the inherent spiritual or instinctual aspects of animals that transcend religious doctrines and ceremonies, connecting them to a fundamental sense of spirituality. Cox articulates primal spirituality as a means of describing glossolalia as an innate spiritual expression that transcends religious dogma and rituals. The notion of glossolalia as primal speech explores a form of preverbal communication akin to that observed in infants. By emphasising the foundational spirituality within Pentecostalism, Cox suggests that glossolalia liberates Christianity from the constraints influenced by Greco-Roman cultural formalities. When applied to non-human creatures, this concept implies that animals possess a direct and unfiltered connection to the Spirit, unencumbered by human-made

[32] Macchia, "Sighs," 69.

[33] Macchia, "Sighs," 69.

[34] For Macchia, the phenomenon of glossolalia is evidence that such transformation has already begun, both among human community and in creation. Macchia, "Sighs," 70. He states, "Here we have an eschatology that incorporates transcendent experience with the realities of our creaturely and historical existence, transforming this existence with the promise of redemption, a promise that includes all of creation." Macchia, "Sighs," 60.

[35] Clough, *On Animals, Systematic Theology*, 41.

structures. For animals, this connection could foster an intimate relationship with God distinct to their non-human nature, as they engage in a free and instinctual form of connection with the Spirit.[36]

This broader understanding of primal expression finds resonance with Moltmann's theology, where he perceives the suffering and cries of creation as deep, Spirit-filled groans that rise to God. For Moltmann, the Spirit not only dwells within but also intercedes for creation's suffering, uniting the voices of oppressed people and the groaning earth in a shared plea for liberation. He writes:

> So today the cries of the oppressed people in the dumb silence of the dying children in the Third World rise up to God. They are the sighs of the Spirit. So from the ruined nature of our earth the sighs of the created beings oppressed and exploited by human beings rise up to God. They are the sighs of the Spirit... the Holy Spirit is related to these cries of the Spirit from the depths, and must take them up; for it is the one same Spirit who in turn interpenetrates the depths of all creation and the depths of the Godhead.[37]

In categorising glossolalia as creaturely sighs and groans, my approach seeks to identify commonalities within the glossolalic experience that can be extended to a broader cross-creaturely context. The objective is to emphasise that, while humans can discern and interpret the transformative potential of glossolalic symbols, the Spirit may convey a spiritual voice by intervening in the innate "sighs and groans" of animal expressions encompassing praise, lament, and the longing for freedom directed towards the Creator. Through a pneumatological perspective, the focus extends beyond the human realm of glossolalia, probing into how the Spirit communicates with and serves as a mediator for non-human creatures.

Glossolalia and the Liberation of Creation

For Macchia, Pentecost and glossolalia establish a significant connection between the language of liberation and unity with the new creation. He argues that Spirit-baptism establishes the framework for the "church's

[36] Harvey Cox, *Fire from Heaven: The Rise of Pentecostal Spirituality and the Reshaping of Religion in the 21st Century* (Boston, MA: a Capo Press, 2001), 81.

[37] Moltmann, *The Spirit of Life*, 77.

identification with Christ's redemptive order."[38] Macchia adds that the Spirit disrupts deeply ingrained prejudices such as classism, sexism, and racism.[39] Additionally, we can include speciesism as another dimension of this disruption for the liberation of a suffering creation. If we consider eschatology as the context for Spirit-baptism in preparation for a renewed creation, it becomes reasonable to argue that unity with our co-creatures should be prioritised in God's divine act in the eschaton.

Viewing creation as a community carries implications for how humans perceive their identity. Humans are intended to engage in reciprocal relationships of giving and receiving, with God as the central focus, recognising that all beings in the evolving world share the intrinsic quality of being creatures. In essence, humans and other animals, despite their distinctions, fundamentally share the same identity as creatures belonging to the same generous God and participating in a unified community. This perspective encourages viewing animals as kin within the Spirit.[40] Moltmann affirms this vision, suggesting that if the Holy Spirit is poured out upon all creation, then the Spirit forms a community of all creatures with God and with one another.[41] This fellowship of creation is one where all beings, each in its own unique way, communicate and connect with each other and with the Creator. As Moltmann puts it, "All creatures speak, even if human beings can no longer hear them. All creatures are aflame with the present glory of God and reflect his glory in a thousand different mirrors."[42]

Traditionally, the idea of Spirit-baptism (often expressed through tongues) is centred on the divine healing of human relationships within the community. However, human prejudices and divisions extend beyond discrimination within one creaturely species. The concept of glossolalia as a transformative experience that influences social relationships could extend to animals and the suffering endured by non-human creatures

[38] Macchia, *Sighs*, 54. See also Cartledge, *Charismatic Glossolalia: An Empirical-Theology Study*, 188. Cartledge describes glossolalia as giving "expression to a longing for God's Kingdom to be consummated here on earth."

[39] Macchia, *Sighs*, 66.

[40] Johnson, "Animals' Praise of God," 267.

[41] Jurgen Moltmann, *God in Creation* (Minneapolis: Fortress Press, 1993), 11.

[42] Moltmann, *The Source of Life*, 134.

due to human actions.[43] If believers receive spiritual gifts to intercede for creation, then the Spirit, as the animating force in animal life, may also intercede on behalf of all creatures through glossolalia, fostering a synergistic relationship that contributes to the liberation of creation in a shared eschaton.[44]

If animals were to express praise and worship to their Creator through the indwelling presence of the Spirit, one of their creaturely functions could involve a desire for *koinonia*, expressed through their glossolalic manifestations of sighs and groans.[45] In this shared solidarity with all creatures, the same Spirit dwelling in the risen Christ intercedes on behalf of creation, serving as a guarantee that the vulnerable future of creation, susceptible to decay, will be preserved and not destroyed.

[43] This would include the effects of global warming in the environment which is attributed to human activities. of global warming.

[44] McLaughlin highlights that Moltmann viewed the Spirit present within "all the breathy sighs in the cosmos, all longing for redemption." McLaughlin, Ryan Patrick. "Anticipating a Maximally Inclusive Eschaton: Jürgen Moltmann's Potential Contribution to Animal Theology." *Journal of Animal Ethics* 4, no. 1 (2014): 25. In light of McLaughlin's reading of Moltmann's theological framework, it could be argued that animals are capable of affective responses to their circumstances or existence, and that the Spirit intercedes by transforming those responses—sighs and groans—into intercessory glossolalia. Building on Green's idea, I propose that it is the Spirit who groans, while the emotional experience is attributed to the one praying, Green, Chris E. *The Crucified God and the Groaning Spirit*, 138. This adds depth to the understanding of how the Spirit mediates animal expressions, further affirming their participation in the broader narrative of creation's longing for redemption.

[45] As Archer states, "creation's participation in God's coming brings about its own liberation." Archer, Kenneth J. *God - Creation's Hope, Creation - God's Home*, 211. Although some of the charisms (1 Cor 12:4–11) are poured out to empower believers in their evangelisation or personal edification, all of the gifts are intended to provide service in order to help build *koinonia*. Kärkkäinen, *Spirit, Reconciliation and Healing in the Community*, 48. The charismata reveal how human existence is never fulfilled in isolation and that God's "creation facilitates humanity's relationship with God. Williams, *Greening the Apocalypse*, 227. Althouse suggests that the charismatic presence of the Spirit presupposes a kenosis of Spirit, as "a sacrifice for the other; and likewise, the Spirit empowers us to be servants for the kingdom, participating with God in his mission to reconcile the whole world." Althouse, Peter. "Implications of the Kenosis of the Spirit for a Creational Eschatology," in *The Spirit Renews the Face of the Earth: Pentecostal Forays in Science and Theology of Creation*, edited by Amos Yong (Eugene, OR: Pickwick, 2009), 171. In terms of animals' participation through prayer and praise, it could be argued that Old Testament animal sacrifice under the Mosaic Law points to one example of the contributions made by animals as participants in God's mission to reconcile the world.

Balaam's Donkey and Glossolalic Speech

What is the significance of animal voice? In Numbers 21:32, a donkey is granted the temporary ability to speak. In this context, God clearly deems the voice of animals important. The story of Balaam's donkey offers a unique expression of intelligible lament from a non-human creature. In this story, God not only utilises a donkey's natural behaviour but also bestows upon her miraculous abilities to serve His purposes.[46] The angel of the Lord is explicitly hostile towards Balaam, warning the prophet that he should only speak the words of Yahweh. In contrast, such animosity is not extended towards the creature, as the donkey also rebukes Balaam. It is evident that the donkey possesses a unique status akin to that of the angel of Yahweh, as both angel and donkey function as agents of God, communicating to Balaam in a mutually complementary manner.[47]

The mechanism of the donkey's temporary faculty of speech directly relates to the admonished human. It is plausible that the miraculous verbal exchange between species points to an Old Testament glossolalic occurrence, where a possible manifestation of either *xenolalia* or *akolalia* could have occurred during this event.[48] This could mean that glossolalia or *xenolalia* is evidenced by the speaker, so that the donkey verbalises a language foreign to her. Alternatively, it could mean that *akolalia* is experienced by the hearer, so that the prophet Balaam can understand the donkey's vocalisation as a comprehensible spoken language.

While this passage could be unpacked further, one important point is clear. In her own emotional protest of ill-treatment, Balaam's donkey

[46] The significance of the biblical narrative where a donkey acquires the momentary faculty of speech in relation to the prophet Balaam, suggests that God "not only employs [the donkey's] natural behaviour for his purposes [but] he also endows her with his supernatural ability for his purposes." Way, Kenneth C. "Animals in the Prophetic World: Literary Reflections on Numbers 22 and 1 Kings 13." *Journal for the Study of the Old Testament* 34, no. 1 (September 2009): 53.

[47] Way, Kenneth C. Animals in the Prophetic World, 54.

[48] According to Yong, *akolalia* refers to a miraculous act of hearing, in which a prophetic message is conveyed to the listener through the vehicle of human language spoken in glossolalic utterance. This message is specific to the listener and is comprehended by them. In contrast, *xenolalia* involves a glossolalic speaker being enabled to speak in a language that is unfamiliar to them. Yong, Amos. *The Truth of Tongues Speech: A Rejoinder to Frank Macchia*, 108–109.

embodies the suffering and distress which torments all non-human creatures, whose grievances generally are restricted to sighs and groans.[49] In contrast to this humanly imposed suffering, we look to Jesus' triumphant entry into Jerusalem. Dignity is bestowed on the donkey ridden by Messiah.[50] Animals, too, find hope in their creaturely Messiah, the Lamb of God.

Implications of Animal Glossolalia

The implications of animal glossolalia are profound in that animals can be seen as part of a prophesying creation.[51] They engage in prophetic expression.[52] Glossolalia, as a cosmic form of communication or ecumenical language of the Spirit, offers a compelling rationale for animals' participation in prophetic discourse.[53] It suggests that the tongues at Pentecost symbolise the first ecumenical language of the church, transcending

[49] In relation to this subject, Klaus Koch performs a brief but interesting study of Hosea 2:21–22. He suggests that the Hebrew verb *anah*, which means "answer," may also have a sense of "to respond in a discussion" and can refer to nonverbal reactions, providing new impulses from one subject to another within a given relationship. Hosea's description of a coherent chain in which God acts in interdependence with earth, creature, human, and nation reflects a "sociology of nature." This indicates that the natural world is a network of interconnected creatures, held together by divine interaction with them. Fretheim, "Nature's Praise of God in the Psalms," 26.

[50] Mark 11:4–7 explains that Messiah's colt is untied, which adds an additional layer to the animal's liberation.

[51] Muers and Griffiths highlight that the theological foundations of non-verbal prophetic expressions broaden our understanding of prophecy, moving beyond an anthropocentric view. They argue that the prophetic voice is not confined to humans but also includes sentient non-human creatures. See Sect. 2.3.4.

[52] For Macchia, the tongues spoken at Pentecost stand as the initial ecumenical language of the church. This first ecumenical language did not derive from Greek or Latin origins. Rather, the tongues manifested at Pentecost underscore the notion that no singular language can assert absolute supremacy when it comes to comprehending truth. These tongues, symbolising unity, encompass the entirety of languages found under the heavens (Acts 2:5). See Frank D Macchia, "Babel and the Tongues of Pentecost: Reversal or Fulfilment? A Theological Perspective," in *Speaking in Tongues: Multi-Disciplinary Perspectives*, ed. Mark J. Cartledge (Eugene, OR: Wipf & Stock Publishers, 2006), 47.

[53] Macchia states, "Spirit baptism implied that the Spirit's redemptive and empowering work is to be experienced and expressed in a vast diversity of tongues that groan for the redemption to come (Rom 8:26), revealing the ultimate inadequacy of human thought and speech to express the divine mystery." Frank D Macchia, "Baptized in the Spirit: Towards a Global Theology of Spirit Baptism," in *The Spirit in the World: Emerging*

human linguistic boundaries and embracing all languages under heaven. That ecumenical language is not restricted to human voices but extends to the non-human realm as well. Just as Spirit baptism manifests through a diversity of tongues, each expressing a longing for redemption, the idea of animal glossolalia highlights the limitations of human language in capturing the fullness of the Spirit's mysteries. This offers a window into the deep connection between the divine, the natural world, and the mysteries that lie beyond human understanding.

Conclusion

Animal glossolalia, emerging from pneumatological imagination, reveals a window into the spiritual life of the non-human world. It makes it possible to see animals not as passive inhabitants of creation, but as dynamic participants in the unfolding divine story—offering their own distinctive voices of praise, lament, and even prophecy. The Spirit, dwelling within creation as the vital force of communication, intercedes for all creatures, weaving their expressions into the cosmic language of communion with the Creator. This inclusive vision shatters anthropocentric boundaries and invites a deeper attunement to the symphony of worship that resounds throughout creation, recognising that animals, too, contribute their voices to the grand chorus of divine praise.

My account in this chapter invites further theological and ethical reflection. For instance, how might we discern the Spirit's work in animal expressions amidst the ambiguity of creation's groaning? While human speech requires discernment to distinguish between divine and deceptive voices, how does this apply to animal vocalisations, including those with apparent social or self-serving functions? Are ordinary vocalisations, such as a cat's meow or a bird's chirp, excluded from the category of glossolalia, or might they simultaneously carry deeper spiritual significance? These questions call for further interdisciplinary research that can hold together the sanctifying work of the Spirit with the complex realities of animal behaviour.

Moreover, this discussion offers a fresh contribution to broader scholarly debates about the nature of language, both human and non-human. How does the liminal nature of human glossolalia intersect with ongoing

Pentecostal Theologies in Global Contexts, ed. Veli-Matti Karkkainen (Grand Rapids, MI: Eerdmans, 2009), 10.

questions about whether non-human communication can be considered "language"? Could glossolalia, understood as a form of cosmic worship, serve as a theological way of bridging these debates? Future research might also explore the limits of this category: whether glossolalia points to a kind of archetypal "speech" accessible to all creatures or to specific Spirit-evoked expressions of divine communication.

CHAPTER 7

The Spirit and the Creatures: Implications and Conclusions

The central contribution of this book is to offer the first sustained exploration of animals through the lens of pneumatology. While theological attention to animals has grown steadily in recent decades, the Spirit has often remained a quiet presence in these conversations—implicit rather than foregrounded. This study brings the Spirit to the centre, weaving together classical and contemporary theological insights into a systematic framework that takes animal life seriously. In doing so, it not only addresses a significant gap in the field but also opens new pathways for theological reflection on creation, redemption, and the Spirit's work among all creatures.

This is also the first large-scale theology of animals written from within a Pentecostal perspective. Pentecostal theology provides fertile ground for this endeavour for several reasons. Its strong pneumatological focus makes it particularly well suited to reimagine the relationship between God, creation, and the Spirit's activity. Its growing ecological consciousness has opened space for animals to be brought more fully into theological conversation. And, perhaps most significantly, engaging animals within Pentecostal theology extends pneumatological reflection into new doctrinal territory, enriching understandings of both animals and the Spirit.

Within the wider landscape of animal theology, this work offers a distinct methodological contribution. While scholars such as David

D. Rizzo, *Animal Pneumatology*, The Palgrave Macmillan Animal Ethics Series, https://doi.org/10.1007/978-3-032-18723-9_7

Clough have laid important systematic foundations through christological approaches, pneumatology has received comparatively little sustained attention. By developing an animal pneumatology, this book demonstrates the theological fruitfulness of attending to the Spirit's role in creation and redemption, and invites further exploration of how pneumatology might reshape the contours of animal theology.

This study speaks into two overlapping conversations. On the one hand, it brings animals more fully into theological focus, affirming them as creatures of inherent worth who stand in relationship to God through Christ. On the other, it offers fresh insights into pneumatology itself, showing how attention to animal life stretches and enriches our understanding of the Spirit's work in creation and redemption. By placing these two fields in conversation, this book opens new pathways for theological reflection, treating animals not only as subjects of concern but as participants in the Spirit's dynamic activity.

To develop this vision, I traced the Spirit's relationship with the non-human world through several key theological themes: creation, predation, salvation, eschatology, and glossolalia. Chapter 2 began with creation, attending to the Spirit's animating breath (*ruach*) and the shared life (*nephesh*) that binds human and non-human creatures together. Here, animals emerged not as background figures in the drama of creation but as co-creatures sustained by the Spirit's presence. Their agency, community life, and relational capacities invite a reimagining of theological frameworks that have too often placed humanity at the centre.

Chapter 3 turned to the thorny issue of natural suffering, especially predation. Rather than trying to resolve the tension between pre-Fall harmony and post-Fall violence, I set these views in dialogue, drawing on Bethany Sollereder's proposal that predation belongs to the fabric of God's good creation. Through a pneumatological lens, predation becomes part of a relational and interdependent world, even as the Spirit holds creation in hope for its ultimate renewal.

From there, Chapter 4 expanded the scope of salvation to include non-human creatures. Drawing on the idea of deep incarnation, I argued that the Spirit's redemptive work encompasses "all flesh," positioning animals as participants—not passive beneficiaries—in God's economy of grace. Chapter 5 carried this vision forward eschatologically, exploring how the Spirit brings all creation into covenantal renewal. Against dualistic models that separate present and future too sharply, I proposed a Spirit-centred vision of cosmic reconciliation in which animals are integral to God's redemptive purposes.

Chapter 6 offered a more imaginative turn, exploring what I call *animal glossolalia*. Here, the vocalisations, movements, and instinctive responses of creatures are reimagined as Spirit-inspired expressions of praise and lament—signs of their participation in the cosmic liturgy. By reframing glossolalia through a pneumatological and creaturely lens, this chapter sought to broaden Pentecostal theology's understanding of worship as something shared across the entire community of creation. Taken together, these chapters develop a theological framework that recognises the Spirit's expansive, inclusive, and life-giving work among non-human creatures. While some questions remain open, the insights gathered here help lay the groundwork for a more comprehensive pneumatology of animals—one that sees the Spirit not only as the sustainer of life but also as the one who draws all creatures into the redemptive and worshipful life of God.

Hearing Creation's Cry: God and the Voices of Animals

Recognising that animals have a voice within the Spirit's dynamic work in creation reshapes both theology and ethics. Animals are not silent ornaments in the backdrop of creation's drama; they are participants whose lives and voices—whether expressed through songs of praise, patterns of communal life, or cries of lament—resonate within the broader purposes of God. The biblical imagination already gestures towards this reality: Psalm 148 summons all creatures into the chorus of praise, while Romans 8 gives voice to creation's groaning as it longs for liberation. These images confront our anthropocentric habits of thought, which have too often pushed non-human creatures to the margins of theological reflection.

To see animals as Spirit-animated beings is to take their presence, their relationships, and even their vocalisations seriously as sites of divine activity. This recognition presses upon us a series of ethical and relational questions. What does it mean to truly hear the voices of non-human creatures? How might our attentiveness to their needs, their vulnerabilities, and even their legal protection be brought into alignment with the justice of God?

Listening to these voices does more than enrich pneumatological discourse; it demands a reorientation of practice. The Spirit's call resounds through the lives of all living creatures, inviting a form of discipleship that is attentive, responsive, and protective. Animals are not merely symbols pointing to divine truths; they are part of the community through which God's Spirit speaks.

"All Flesh" Baptising Work of the Spirit

The Spirit's redemptive work, poured out upon "all flesh," draws animals unmistakably into the wider economy of salvation. Romans 8 offers a vision of redemption that refuses to be confined to human experience alone: creation groans in chorus, straining towards the moment when liberation will arrive in full. Animals, as sentient beings who share in the vulnerabilities and sufferings of a fractured world, are not bystanders to this drama but participants within it. Their lives, too, are swept up in the Spirit's redemptive movement, which breathes hope into every corner of creation.

This vision invites us to rethink salvation not as a narrowly human affair but as a cosmic reality animated by the Spirit—one that integrates the destinies and flourishing of all creatures. To speak of "all flesh" is to affirm animals as co-creatures who stand within the unfolding purposes of God as partners in the divine economy, not simply as instruments for human use or ecological props within a grander human narrative.

Such a theological shift unsettles the familiar paradigm of dominion. It beckons us towards an ethic of partnership, where our relationships with animals mirror the Spirit's unifying and inclusive activity. At the eschatological horizon, this vision stretches forward to the day when the Spirit's work reaches its fullness, reconciling and flourishing all creation—human and non-human alike.

Reconciliation and the Animal

The Spirit's breath (*ruach*), which animates all life, serves as a unifying force that binds human and non-human creatures within the dance of creation. This pneumatological vision reveals creation as fundamentally relational: all life, sustained by the same animating breath, stands together as recipients of the Spirit's sustaining presence. Within this framework, reconciliation is no longer an exclusively human experience but a reality that encompasses the entire community of creation. Animals, like humans, are drawn into the Spirit's redemptive movement, bound into the same divine rhythm of life and renewal.

The concept of deep incarnation deepens this vision still further. Christ's assumption of flesh in the incarnation is understood not merely as a solidarity with humanity but as a solidarity with all embodied life. This identification with creation situates animals within the redemptive scope of the incarnation, affirming their value and their participation in the divine narrative of restoration. Deep incarnation reframes reconciliation as more than the mending of human relationships—it extends to the healing of creation itself, to the restoration of all relationships between creatures and their Creator.

Eschatology gives this vision its horizon. The Spirit's work of renewal reaches towards a future in which creation is reconciled and made whole, where fractures give way to harmony and every living thing finds its place in God's purposes. The biblical imagination offers glimpses of this future: Isaiah's vision of the peaceable kingdom, where predator and prey coexist in shalom; Revelation's cosmic worship, where all creatures lift their voices in praise. In these scenes, the boundaries that divide human and non-human fade. Animals stand not at the margins but within the centre of God's eschatological vision, participating fully in the flourishing life of the new creation.

Theological Implications

The pneumatological vision developed in this book moves beyond the well-trodden terrain of human relationships and Trinitarian dynamics to embrace the non-human world as an essential theological partner. By framing animals as co-creatures whose lives are animated and sustained by the Spirit, I offer a foundation for reimagining humanity's posture towards them. This perspective does not rest merely on arguments for animal rights; rather, it is grounded in the theological recognition of the Spirit's deeply relational activity in animal life. Animals emerge not as peripheral to God's purposes but as participants drawn into the broader drama of redemption and renewal through the Spirit's validating presence.

I contend that the Spirit's activity is not limited to human life but extends to the entire community of creation. This widening of the pneumatological lens invites Pentecostal theology to recognise animals as co-recipients of the Spirit's life-giving presence and activity. Such a shift not only broadens the scope of Pentecostal pneumatology but also foregrounds the ethical and relational responsibilities of the Pentecostal movement towards non-human creatures.

For Pentecostal scholarship, this presents both a challenge and an opportunity. It calls for deeper engagement with the theological and practical implications of including animals within the Spirit's redemptive vision. This engagement could lay the groundwork for a distinct Pentecostal animal ethic—one capable of informing and transforming practical ministry. With Pentecostal communities flourishing in regions most affected by ecological degradation, a Spirit-shaped ethic that includes animals within a broader creation care framework is not simply timely; it is urgent. Pentecostalism's dynamic theology of the Spirit offers unique resources for articulating such an ethic, one that leans into the interdependence of all creation under the Spirit's care.

This proposal also resonates with Pentecostalism's historical legacy of aligning theological conviction with social justice. Just as Pentecostal theology once stood alongside the Civil Rights Movement, so too could it rise to confront other forms of oppression, including speciesism.[1] Drawing from its rich tradition of advocating for justice and equality, Pentecostalism is well-positioned to extend its witness to include the care and protection of animals. Such a move would not only reflect the Spirit's expansive work in creation but also offer a robust theological and ethical response to the ecological crises shaping our time.[2]

[1] Swoboda also makes this connection, see A. J. Swoboda, "Looking the Wrong Way: Salvation and the Spirit in Pentecostal Eco-Theology," in *A Liberating Spirit: Pentecostals and Social Action in North America*, ed. Michael Wilkinson and Steven M Studebaker (Eugene, OR: Pickwick Publications, 2010), 240 cf:40.

[2] Walter Hollenweger's vision for a theology of development and social justice, rooted in Paul's understanding of the *charismata*, offers a powerful framework for expanding Pentecostal theology to encompass a broader, more inclusive vision of creation care. By magnifying the Spirit's work within diverse cultural contexts, Hollenweger highlights that the Creator Spirit is actively present in the unique gifts and expressions of every community. These gifts, nurtured and expressed within their cultural frameworks, contribute to the common good of both community and society. W. J. Hollenweger, "Creator Spiritus: The Challenge of Pentecostal Experience to Pentecostal Theology," *Theology* 81, no. 679 (January 1, 1978): 40. This call to recognise the Spirit's work in cultures, even when expressed in unfamiliar ways such as "strange tongues," "foreign dances," or "oral categories," invites a further broadening of theological imagination. Just as Hollenweger urges attention to the Spirit's activity within diverse human communities, I suggest that such a theology of development and social justice must also extend to include non-human animals as participants in the Spirit's creative and redemptive work. A truly charismatic

Integrating animal ethics into Pentecostal theology has the potential to ignite a new wave of Spirit-led activism, animated by the conviction that the Spirit's transformative presence permeates all of life. This reorientation would both enrich Pentecostal theology and equip the movement to address ecological and ethical challenges with renewed depth and urgency.

This book contributes to that trajectory by advancing theological concepts such as *animal glossolalia* and the Spirit-filled participation of animals in cosmic worship. It invites Pentecostal theology to widen its understanding of Spirit-filled expression beyond human experience, recognising the voices and lives of non-human creatures as integral to the Spirit's work. In doing so, it presses Pentecostals to rethink their doctrines and practices in light of a pneumatological vision that is radically inclusive. Emphasising animals' participation in praise, lament, prophecy, and redemption resonates deeply with Pentecostalism's core emphases on dynamic worship and Spirit-empowered engagement with the world.

Such a vision not only enriches Pentecostal theology but also opens new avenues for justice, care, and ecological stewardship. By recovering its historical commitment to Spirit-led transformation and applying it to the more-than-human world, Pentecostalism can offer a distinctive and theologically rich contribution to contemporary ecological discourse.

Moreover, this perspective provides fertile ground for interfaith dialogue, particularly with traditions that affirm the sacredness of animals, such as Indigenous spiritualities. Shared commitments to care, respect, and interdependence create space for mutual learning and collective action in response to ecological degradation. By situating animals as

social theology—one that values the *charismata* as expressions of the Spirit's presence—cannot neglect the voices, lives, and communities of animals that also bear witness to the Spirit's animating and sustaining power. By integrating animals into this vision, Pentecostal theology could develop a distinctive approach to social justice that includes compassionate conservation and ecological responsibility as integral to its mission. The *charismata* might be understood not only as gifts expressed through human culture but also as reflective of the Spirit's work in animals. Such a theology, following Hollenweger's lead, would challenge Pentecostal communities to embrace a more inclusive understanding of mission—one that celebrates the Spirit's work not only in diverse human cultures but also in the lives and habitats of non-human creatures. This would affirm the dignity of all creation and align Pentecostal theology more closely with the Spirit's expansive and transformative work in the world. By attending to the interconnected flourishing of all life, Pentecostal theology could embody a charismatic vision that resonates deeply with the Creator Spirit's call for justice, care, and reconciliation across the whole of creation.

Spirit-filled beings, this pneumatological vision offers common theological ground upon which interfaith initiatives can address shared ecological and ethical responsibilities.

Further Implications and Trajectories

Future explorations of a more holistic theology of animals would do well to learn deeply from Indigenous perspectives. Indigenous worldviews consistently recognise animals as co-creatures, participating in the sacred balance of creation. Attending to these perspectives offers a crucial corrective to entrenched Western dualisms that separate humanity from the rest of creation. By bringing Indigenous knowledge into conversation with theological reflection, new pathways open for discerning the Spirit's work in creation—pathways that foster reconciliation between theological traditions and Indigenous wisdom.

This widening of dialogue does not end there. The pneumatological vision developed in this book also lends itself to fruitful engagement with other Christian traditions, particularly Orthodox and Catholic theologies. Orthodox theology's emphasis on the Spirit as the giver of life and Catholic sacramental perspectives that perceive creation as relational and imbued with sacred presence resonate strongly with a pneumatology attentive to animals. Entering into dialogue with these traditions could deepen and broaden theological reflection, integrating animal pneumatology into wider sacramental and eschatological frameworks. Such conversations promise to enrich not only animal theology but also the shared theological imagination of the Church as it contemplates the Spirit's life-giving presence across all creation.

Unresolved Questions and Future Directions in Animal Pneumatology

Several theological issues within animal pneumatology warrant further exploration. For instance, the ontological and eschatological implications of the Spirit's work in animals need a more robust systematic treatment. How do we account for the Spirit's sustaining presence amidst the suffering and predation experienced by animals? Furthermore, how might eschatological hope address the realities of animal extinction and environmental degradation? These questions intersect with broader concerns in theodicy.

Another unresolved issue pertains to the relationship between animals and sacramentality. How might the sacredness of animals be articulated in relation to traditional theological symbols and sacraments? Could animals themselves embody sacramental significance as signs of God's presence and grace within creation? Developing a pneumatology of animals that addresses these questions would deepen theological engagement with the non-human world and enrich the discourse on the Spirit's work within creation.

In the realm of practical theology, there may be merit in developing ecclesial praxis to explore the healing and comforting presence of animals in pastoral ministry. Recognising the Spirit's relational and redemptive presence in animal life invites churches to reflect on how they might honour the dignity of animals through their liturgical, ethical, and communal practices. Developing resources for Pentecostal churches such as liturgies that honour animals, sermons advocating for animal rights and welfare, and community programs that address compassionate conservation could translate these theological insights into meaningful and actionable ministry. Even simple, practical shifts can carry deep significance. For example, replacing traditional catering and food hampers with vegetarian or vegan alternatives signals a commitment to the flourishing of all creatures. Initiatives such as these allow ministries to embody their theology in everyday acts, witnessing to the Spirit's reconciling work not only through proclamation but through shared practices that nurture life. Such practices hold the potential to enrich Pentecostal engagement with ecological concerns and foster a more distinctive approach to pastoral care.

Recognising the role of animals in supporting individuals with disability, illness, grief, and depression offers opportunities to consider ministry as a creaturely inclusive space. Initiatives such as the inclusion of comfort animals in church settings or pastoral visits could provide tangible expressions of the Spirit's reconciling and nurturing presence. Furthermore, this vision could extend to reimagining church architecture, where buildings are designed to coexist with and even serve local wildlife, such as providing sanctuaries for koala populations. These approaches not only embody the Spirit's work in the reconciliation of creation but also affirm animals as integral to the Church's mission.

By addressing the human-animal divide and embracing animals as co-participants in the Spirit's work, such practices reflect the church's responsibility to engage in a ministry that is both theologically sound and ecologically conscious. Grounding pastoral ministry in a holistic understanding of creation care invites ecumenical collaboration that moves the church beyond human spaces into wild places, where a renewed commitment to ethical and relational dimensions of theology can resonate deeply in an era of ecological crisis. With pneumatological considerations of the non-human animal remaining nascent, such initiatives may offer rich and important insights for the church and academy.

At the end of a monograph on animal theology, it is worth remembering that theology itself exists under the sign of the animal, as exemplified in Balaam's donkey—a creature who cried out to be heard, embodying a prophetic voice overlooked by human awareness. Perhaps, as *vox clamantis in deserto* ("a voice crying in the wilderness"), the animal continues to call theology to a deeper self-examination, challenging it to listen attentively and respond faithfully to the Spirit's work in creation.

Bibliography

Almond, Rosamunde. "Living Planet Report 2022: Building a Nature-Positive Society." Gland: WWF, 2022.

Althouse, Peter. "Implications of the Kenosis of the Spirit for a Creational Eschatology: A Pentecostal Engagement with Jürgen Moltmann." In *The Spirit Renews the Face of the Earth: Pentecostal Forays in Science and Theology of Creation*, edited by Amos Yong, 155–72. Eugene, OR: Pickwick Publications, 2009.

Althouse, Peter. *Spirit of the Last Days: Pentecostal Eschatology in Conversation with Jurgen Moltmann*. Journal of Pentecostal Theology: Suppliment Series 25. Edinburgh, UK: T&T Clark, 2003.

Ambrose, Saint. *On the Mysteries and The Treatise on the Sacraments*. Aeterna Press, 2015.

Andrews, Kristin. *The Animal Mind: An Introduction to the Philosophy of Animal Cognition*. London: Routledge, 2020.

Archer, Kenneth J. "God—Creation's Hope, Creation—God's Home: Pentecostal Theological Response to Terence E. Fretheim's God and World in the Old Testament: A Relational Theology of Creation." *Journal of Pentecostal Theology* 19, no. 2 (2010): 198–212. https://doi.org/10.1163/174552510X526214.

Astley, Jeff. "Evolution and Evil: The Difference Darwinism Makes in Theology and Spirituality." In *Reading Genesis After Darwin*, edited by Stephen C Barton and David Wilkinson, 163–80. Oxford, UK: Oxford University Press, 2009.

Atkins, Peter Joshua. "Praise by Animals in the Hebrew Bible." *Journal for the Study of the Old Testament* 44, no. 3 (March 2020): 500–513.

D. Rizzo, *Animal Pneumatology*, The Palgrave Macmillan Animal Ethics Series, https://doi.org/10.1007/978-3-032-18723-9

Audlin, James David. "Breathing the Name of God: 'YHWH' and 'Elohim' in the Gospel of John." In *The Gospel of John Restored and Translated*, edited by Volcán Barú, Vol. 2. CreateSpace, 2013.

Balabanski, Vicky. "Revisiting the 'Voice of Earth.' Earth Bible Principle #3." *Journal of Ethics in Antiquity and Christianity* 5 (2023): 75–86. https://doi.org/10.25784/jeac.vSil.1204.

Baron, Eugene, and Amos Yong, eds. *Pentecostal Missiology and Environmental Degradation.* Carlisle, UK: Langham Global, Forthcoming.

Barth, Karl. *Church Dogmatics.* Vol. 1/2. Edinburgh: T&T Clark, 1956.

Bauckham, Richard. *Bible and Ecology: Rediscovering the Community of Creation.* London: Darton, Longman and Todd, 2010.

Bauckham, Richard. "Creation's Praise of God in the Book of Revelation." *Biblical Theology Bulletin* 38, no. 2 (May 1, 2008): 55–63. https://doi.org/10.1177/014610790803800202O1.

Bauckham, Richard. "Joining Creation's Praise of God." *Ecotheology: Journal of Religion, Nature & the Environment* 7, no. 1 (July 2002): 45.

Bauckham, Richard. *Living with Other Creatures: Green Exegesis and Theology.* Wako, TX: Baylor University Press, 2011.

Bekoff, Marc. *The Emotional Lives of Animals: A Leading Scientist Explores Animal Joy, Sorrow, and Empathy—and Why They Matter.* Novato: New World Library, 2008.

Bekoff, Marc. "The Evolution of Animal Play, Emotions, and Social Morality: On Science, Theology, Spirituality, Personhood, and Love." *Zygon: Journal of Religion & Science* 36, no. 4 (December 2001): 615. https://doi.org/10.1111/0591-2385.00388.

Benz-Schwarzburg, Judith. *Cognitive Kin, Moral Strangers? Linking Animal Cognition, Animal Ethics & Animal Welfare.* Human-Animal Studies 43. Leiden, The Netherlands: Brill, 2020. https://brill-com.alphacrucis.idm.oclc.org/display/title/35289.

Bergmann, Sigurd. *Creation Set Free: The Spirit as Liberator of Nature.* Grand Rapids, MI: Eerdmans, 2005.

Bergmann, Sigurd. "Where on Earth Does the Spirit 'Take Place' Today? Considerations on Pneumatology in the Light of the Global Environmental Crisis." In *Christian Faith and the Earth : Current Paths and Emerging Horizons in Ecotheology.*, edited by Ernst M. Conradie and editor of compilation., 51–64. New York: Bloomsbury T&T Clark, 2014.

Bonhoeffer, Dietrich. *Creation and the Fall: A Theological Interpretation of Genesis 1–3.* Translated by John C Fletcher. London, UK: SCM Press, 1959.

Brooks, Adrian J. "Breathing Forth the Word: Yves Congar's Articulation of the Activity of the Holy Spirit in the Life of Christ." *New Blackfriars* 101, no. 1092 (2020): 196–205.

Brooks Pribac, Teja. "Spiritual Animal: A Journey into the Unspeakable." *Journal for the Study of Religion, Nature and Culture* 11, no. 3 (2017): 340–60. https://doi.org/10.1558/jsrnc.31519.

Cadwallader, Alan H. "Earth as Host or Stranger?: Reading Hebrews 11 from Diasporan Experience." In *The Earth Story in the New Testament*, edited by Norman C. Habel and Vicky Balabanski, 148–79. London/ New York: Sheffield Academic Press, 2002.

Cartledge, Mark J. *Charismatic Glossolalia: An Empiracle-Theology Study*. England: Ashgate Publishing Company, 2002.

Chabout, Jonathan, Abhra Sarkar, David B. Dunson, and Erich D. Jarvis. "Male Mice Song Syntax Depends on Social Contexts and Influences Female Preferences." *Frontiers in Behavioral Neuroscience* 9 (April 1, 2015): 76. https://doi.org/10.3389/fnbeh.2015.00076.

Charles, Nickie. "'Animals Just Love You as You Are': Experiencing Kinship across the Species Barrier." *Sociology* 48, no. 4 (2014): 715–30.

Clark, Mary T. *Augustine*. Outstanding Christian Thinkers. London: Geoffrey Chapman, 1994.

Clifton, Shane. "Editorial: Identity and the Shape of Pentecostal Theology." *Australasian Pentecostal Studies* 19 (June 15, 2017): 4–23.

Clifton, Shane. "Preaching the 'Full Gospel' in the Context of Global Environmental Crisis." In *The Spirit Renews the Face of the Earth: Pentecostal Forays in Science and Theology of Creation*, edited by Amos Yong, 117–34. Eugene, OR: Pickwick Publications, 2009.

Clough, David L. *On Animals, Systematic Theology*. London: T&T Clark, 2014.

Clough, David L. *On Animals, Theological Ethics*. Vol. 2. London: T&T Clark, 2019.

Coakley, Sarah, ed. "The Charismatic Constituency: Embarrassment or Riches?" In *God, Sexuality, and the Self: An Essay "On the Trinity,"* 152–89. Cambridge: Cambridge University Press, 2013. https://doi.org/10.1017/CBO9781139048958.006.

Conradie, Ernst. "Do Only Humans Sin? In Conversation with Frans de Waal." In *Issues in Science and Theology: Are We Special?: Human Uniqueness in Science and Theology*, edited by Michael Fuller, Dirk Evers, Anne Runehov, and Knut-Willy Sæther, 117–32. Issues in Science and Religion: Publications of the European Society for the Study of Science and Theology. Springer, 2017.

Conradie, Ernst. *The Earth in God's Economy: Creation, Salvation and Consummation in Ecological Perspective*. Zürich: Lit Verlag, 2015.

Conradie, Ernst. "On Social Evil and Natural Evil: In Conversation with Christopher Southgate." *Zygon: Journal of Religion & Science* 53, no. 3 (September 2018): 752–65. https://doi.org/10.1111/zygo.12427.

Conradie, Ernst. *Redeeming Sin? : Social Diagnostics amid Ecological Destruction.* Religious Ethics and Environmental Challenges Series. Lanham: Lexington Books/Fortress Academic, 2017.

Conradie, Ernst. *Secular Discourse on Sin in the Anthropocene : What's Wrong with the World?* Environment and Society Series. Lanham: Lexington Books/ Fortress Academic, 2020.

Conradie, Ernst. "What Is the Place of the Earth in God's Economy? Doing Justice to Creation, Salvation and Consummation." In *Christian Faith and the Earth : Current Paths and Emerging Horizons in Ecotheology*, edited by Ernst M. Conradie, 65–96. New York: Bloomsbury T&T Clark, 2014.

Cox, Harvey. *Fire from Heaven: The Rise of Pentecostal Spirituality and the Reshaping of Religion in the 21st Century.* Boston, MA: Capo Press, 2001.

Cunningham, David S. "The Way of All Flesh: Rethinking the Imago Dei." In *Creaturely Theology: On God, Humans and Other Animals*, edited by Celia Deane-Drummond and David Clough, 100–117. London: SCM Press, 2009.

Cunningham, Paul. "The Case for Animal Spirituality-Part 1: Conceptual Challenges, Methodological Considerations, and the Question of Animal Consciousness." *Journal for the Study of Religion, Nature & Culture* 16, no. 2 (April 2022): 186–224. https://doi.org/10.1558/jsrnc.18801.

Cunningham, Paul. "The Case for Animal Spirituality-Part 2: Logical Arguments, Empirical Evidence, and Practical Consequences for Human Society." *Journal for the Study of Religion, Nature & Culture* 16, no. 2 (April 2022): 225–63. https://doi.org/10.1558/jsrnc.18802.

Davis, Anita. "A Pentecostal Theological Perspective on Ecofeminism." *Journal of Pentecostal and Charismatic Christianity* 44, no. 1 (2024): 54–69.

Davis, Anita. "Pentecostal Approaches to Ecotheology: Reviewing the Literature." *Australasian Pentecostal Studies* 22, no. 1 (June 8, 2021): 4–33.

Davis, Anita, and Jacqueline Grey. "The Apocalyptic Destruction of the Earth in Isaiah 24 and Its Place in a Pentecostal Ecotheology." In *Pentecostal Missiology and Environmental Degradation*, edited by Eugene Baron and Amos Yong. Carlisle, UK: Langham Global, Forthcoming.

De Waal, Frans B. M. *Good Natured: The Origins of Right and Wrong in Humans and Other Animals.* Cambridge: Harvard University Press, 1996.

De Waal, Frans B. M., and Kristin Andrews. "The Question of Animal Emotions." *Science* 375, no. 6587 (2022): 1351–52. https://doi.org/10.1126/science.abo2378.

De Waal, Frans B. M., and Pier Francesco Ferrari. "Towards a Bottom-up Perspective on Animal and Human Cognition." *Trends in Cognitive Sciences* 14, no. 5 (May 2010): 201–7. https://doi.org/10.1016/j.tics.2010.03.003.

De Waal, Frans B.M. "Natural Normativity: The 'is' and 'Ought' of Animal Behavior." *Behaviour* 151, no. 2/3 (2014): 185–204.

Deane-Drummond, Celia. *Christ and Evolution: Wonder and Wisdom.* Minneapolis: Fortress Press, 2009.

Deane-Drummond, Celia. *Ecology in Jürgen Moltmann's Theology*. Eugene, OR: Wipf & Stock Publishers, 2016.

Deane-Drummond, Celia. *The Wisdom of the Liminal: Evolution and Other Animals in Human Becoming*. Grand Rapids, MI: Eerdmans, 2014.

Deverell, Garry Worete. *Contemplating Country: More Gondwana Theology.* Eugene, OR: Wipf and Stock Publishers, 2023.

Dillard, Anne. *Pilgrim at Tinker Creek.* New York: Harper Perennial Modern Classics, 2013.

Dyer, Keith D. "When Is the End Not the End? The Fate of the Earth in Biblical Eschatology (Mark 13)." In *The Earth Story in the New Testament*, edited by Norman C. Habel and Vicky Balabanski, 45–75. London/ New York: Sheffield Academic Press, 2002.

Eaton, Matthew. "Ecocide as Deicide: Eschatological Lamentation and the Possibility of Hope." In *Integral Ecology for a More Sustainable World: Dialogues with Laudato Si'*, edited by Dennis O'Hara, Matthew Eaton, and Michael T. Ross. London, UK: Lexington Books, 2019.

Eaton, Matthew. *Incarnate Earth: Deep Incarnation and the Face of Christ.* London: Routledge, 2023. https://doi.org/10.4324/9781003287346.

Edwards, Denis. *Breath of Life- A Theology of Creator Spirit*. Maryknoll, NY: Orbis Books, 2004.

Edwards, Denis. *Deep Incarnation: God's Redemptive Suffering with Creatures.* Maryknoll, NY: Orbis Books, 2019.

Edwards, Denis. *How God Acts: Creation, Redemption and Special Divine Action.* MPLS: Fortress Press, 2010.

Edwards, Denis. "Redemption of Animals in an Incarnational Theology." In *Creaturely Theology: On God, Humans and Other Animals*, edited by Celia Deane-Drummond and David Clough. London: SCM Press, 2009.

Falk, Dan. "Insects and Other Animals Have Consciousness, Experts Declare." Quanta Magazine, April 19, 2024. https://www.quantamagazine.org/insects-and-other-animals-have-consciousness-experts-declare-20240419/.

Fiddes, Paul S. *The Creative Suffering of God*. Oxford: Clarendon Press, 1991.

Fretheim, Terence E. *God and World in the Old Testament*. Nashville: Abingdon Press, 2005.

Fretheim, Terence E. "Nature's Praise of God in the Psalms." *Ex Auditu* 3 (1987): 16–30.

Furst, Jill Leslie McKeever. *The Natural History of the Soul in Ancient Mexico.* New Haven and London: Yale University Press, 1995.

Garvey, Jon. *God's Good Earth: The Case for An Unfallen Creation*. Eugene, OR: Cascade Books, 2019.

Gloveli, Natalie, Jean Simonnet, Wei Tang, Miguel Concha-Miranda, Eduard Maier, Anton Dvorzhak, Dietmar Schmitz, and Michael Brecht. "Play and Tickling Responses Map to the Lateral Columns of the Rat Periaqueductal Gray." *Neuron* 111, no. 19 (October 4, 2023): 3041–3052.e7. https://doi.org/10.1016/j.neuron.2023.06.018.

Goodall, Jane. "Do Chimpanzees Have Souls?: Possible Precursors of Religious Behavior in Animals." In *Spiritual Information: 100 Perspectives on Science and Religion*, edited by Charles L. Harper and Charles L. Harper Jnr, 275–78. West Conshohocken: Templeton Press, 2005.

Green, Chris E. "The Crucified God and the Groaning Spirit: Toward a Pentecostal Theologia Crucis in Conversation with Jürgen Moltmann." *Journal of Pentecostal Theology* 19, no. 1 (April 2010): 127–42. https://doi.org/10.1163/174552510X489946.

Gregersen, Niels Henrik. "The Cross of Christ in an Evolutionary World." *Dialog: A Journal of Theology* 40, no. 3 (September 2001): 192. https://doi.org/10.1111/0012-2033.00075.

Griffiths, John D. "Spirit-Baptised Creation: Locating Pentecost in the Meta-Narrative of Creation and Its Implications for a Pentecostal Ecology." *Australasian Pentecostal Studies* 22, no. 1 (June 9, 2021): 46–60.

Griffiths, John D. "Upon All Flesh: An Ecological Reading of Acts 2.17–21." *Journal for the Study of the New Testament*, Forthcoming.

Habel, Norman C. *Readings from the Perspective of Earth*. Sheffield, England: Sheffield Academic Press, 2000.

Hausoul, Raymond R. *God's Future for Animals: From Creation to New Creation*. Eugene, OR: Wipf and Stock, 2021.

Hiebert, Theodore. *The Yahwist's Landscape: Nature and Religion in Early Israel*. Minneapolis: Fortress Press, 2008.

Hollenweger, W. J. "Creator Spiritus: The Challenge of Pentecostal Experience to Pentecostal Theology." *Theology* 81, no. 679 (January 1, 1978): 32–40. https://doi.org/10.1177/0040571X7808100106.

Hollenweger, W. J. "Creator Spiritus: The Challenge of Pentecostal Experience to Pentecostal Theology." *Theology* 81, no. 679 (January 1, 1978): 32–40.

Horrell, David G, Cherryl Hunt, and Christopher Southgate. *Greening Paul: Rereading the Apostle in a Time of Ecological Crisis*. Waco, Texas: Baylor University Press, 2010.

Humphreys, Rebekah. *Animals, Ethics and Language: The Philosophy of Meaningful Communication in the Lives of Animals*. The Palgrave Macmillan Animal Ethics Series. Switzerland: Palgrave Macmillan, 2023.

Jardat, Plotine, Océane Liehrmann, and et.al. "Horses Discriminate between Human Facial and Vocal Expressions of Sadness and Joy." *Animal Cognition* Pre-Print (May 2023). https://doi.org/10.21203/rs.3.rs-2859848/v1.

Jenkins, Willis. "Loving Swarms: Religious Ethics amid Mass Extinction." In *Extinction and Religion*, edited by Jeremy H. Kidwell and Stefan Skrimshire, 16–59. Religion and the Human Series. Bloomington, Indiana: Indiana University Press, 2024.

Jensen, David H. "Discerning the Spirit: A Historical Introduction." In *The Lord and Giver of Life: Perspectives on Constructive Pneumatology*, edited by David H Jensen. Louisville; London: Westminster John Knox Press, n.d.

Johnson, Elizabeth. *Women, Earth, and Creator Spirit*. New York: Paulist Press, 1993.

Johnson, Elizabeth. "Animals' Praise of God." *Interpretation* 73, no. 3 (July 2019): 259–71. https://doi.org/10.1177/0020964319838804.

Johnson, Elizabeth. *Ask the Beasts: Darwin and the God of Love*. London, UK: Bloomsbury Continuum, 2014.

Jones, Ian Charles. "Dominion and Communion: Patristic Theology and the Ethics of Humanity's Relationship with Animal Creation." Ph.D., Fordham University, 2016. https://search.proquest.com/religion/docview/1780959752/abstract/D5B8DA87705E46F5PQ/4.

Kamenicky, Michael Austin. "'You Will Not Surely Die': The Pentecostal Aesthetics and Ethics of Serpent Handling." *The Journal of Religious Ethics* 52, no. 2 (2024): 189–208. https://doi.org/10.1111/jore.12469.

Kärkkäinen, Veli-Matti. *Creation and Humanity: A Constructive Christian Theology for the Pluralistic World, Volume 3*. Illustrated edition. Grand Rapids, MI: Eerdmans, 2015.

Kärkkäinen, Veli-Matti. "Spirit, Reconciliation and Healing in the Community: Missiological Insights from Pentecostals." *International Review of Mission* 94, no. 372 (January 2005): 43–50.

Kärkkäinen, Veli-Matti. *Toward a Pneumatological Theology: Pentecostal and Ecumenical Perspectives on Ecclesiology, Soteriology, and Theology of Mission*. Lanham,Maryland: University Press of America, 2002.

Kemmerer, Lisa. "Theorizing 'Others.'" In *Theorizing Animals: Rethinking Humananimal Relations*, edited by Nik Taylor and Tania Signal, 87–106. Leiden: Brill, 2011.

Kidwell, Clara Sue, Homer Noley, and George E. "Tink" Tinker. *A Native American Theology*. Maryknoll, N.Y: Orbis Books, 2001.

King, Gerald W. "Evolving Paradigms: Creationism as Pentecostal Variation on a Fundamentalist Theme." In *The Spirit Renews the Face of the Earth: Pentecostal Forays in Science and Theology of Creation*, edited by Amos Yong, 93–114. Eugene, OR: Pickwick Publications, 2009.

Kosuke Koyama. *Water Buffalo Theology*. Maryknoll: Orbis Books, 1999.

Lamp, Jeffrey S. "Incarnate Pneumatology: Thinking Ecologically about the Spirit." Presented at the 45th Annual Meeting of the Society for Pentecostal Studies, 2016.

Lessard, Michael. "Beyond Sentience: Legally Recognizing Animals' Sociability and Agency." *Journal of Animal Ethics* 14, no. 1 (2024): 89–109.

Levison, John R. *Filled with the Spirit*. Grand Rapids, MI: Eerdmans, 2009.

Linzey, Andrew. *Animal Gospel*. Louisville, KY: Westminster John Knox Press, 1998.

Linzey, Andrew. *Animal Theology*. Champaign, IL: University of Illinois Press, 1995.

Linzey, Andrew. *Christianity and the Rights of Animals*. Eugene, OR: Wipf and Stock, 2016.

Linzey, Andrew. *Creatures of the Same God: Explorations in Animal Theology*. New York: Lantern Books, 2009.

Linzey, Andrew. *Why Animal Suffering Matters: Philosophy, Theology, and Practical Ethics*. Oxford: Oxford University Press, 2013.

Linzey, Andrew, and Dan Cohn-Sherbok. *After Noah: Animals and the Liberation of Theology*. London: Mowbray, 1997.

Linzey, Andrew, and Dorothy Yamamoto, eds. *Animals on the Agenda: Questions about Animals for Theology and Ethics*. Champaign, IL: University of Illinois Press, 2010.

Linzey, Clair. *Developing Animal Theology: An Engagement with Leonardo Boff*. London: Routledge, 2022.

Lloyd, Michael. "Are Animals Fallen?" In *Animals on the Agenda*, edited by Andrew Linzey and Dorothy Yamamoto, 147–60. London, UK: SCM Press, 1998.

Macchia, Frank D. "Babel and the Tongues of Pentecost: Reversal or Fulfilment? A Theological Perspective." In *Speaking in Tongues: Multi-Disciplinary Perspectives*, edited by Mark J. Cartledge. Eugene, OR: Wipf & Stock Publishers, 2006.

Macchia, Frank D. *Baptized in the Spirit: A Global Pentecostal Theology*. Grand Rapids, MI: Zondervan, 2006.

Macchia, Frank D. "Baptized in the Spirit: Towards a Global Theology of Spirit Baptism." In *The Spirit in the World: Emerging Pentecostal Theologies in Global Contexts.*, edited by Veli-Matti Karkkainen. Grand Rapids, MI: Eerdmans, 2009.

Macchia, Frank D. *Jesus the Spirit Baptizer: Christology in Light of Pentecost*. Grand Rapids, MI: Eerdmans, 2018.

Macchia, Frank D. "Sighs Too Deep for Words: Toward a Theology of Glossolalia." *Journal of Pentecostal Theology* 1, no. 1 (1992): 47–73.

Macchia, Frank D. "The Spirit of Life: Toward a Creation Pneumatology." In *Third Article Theology: A Pneumatological Dogmatics*, edited by Myk Habets, 113–30. Minneapolis: Fortress Press, 2016.

Maier, Harry O. "Green Millennialism: American Evangelicals, Environment and the Book of Revelation." In *Ecological Hermeneutics Biblical, Historical and*

Theological Perspectives, edited by David G. Horrell, Cherryl Hunt, Christopher Southgate, and Stavrakopoulou, 246–65. London/ New York: T&T Clark, 2010.

McGinn, Sheila E. "All Creation Groans in Labour: Paul's Theology of Creation in Romans 8:18–23." In *Earth, Wind and Fire: Biblical and Theological Perspectives on Creation*, edited by Carol J Dempsey and Mary Margaret Pazdan, 114–23. Collegeville, Minnesota: Liturgical Press, 2004.

McGraw, John J. "Tongues of Men and Angels: Assessing the Neural Correlates of Glossolalia." In *Religion and the Body*, edited by David Cave and Rebecca Sachs Norris, 57–80. Modern Science and the Construction of Religious Meaning. Brill, 2011. https://www.jstor.org/stable/10.1163/j.ctv2gjx03h.7.

McLaughlin, Ryan Patrick. "Andrew Linzey (1952–): Animal Theology." In *Animal Theologians*, edited by Andrew Linzey and Clair Linzey, 422–39. NY: Oxford University Press, 2023.

McLaughlin, Ryan Patrick. "Anticipating a Maximally Inclusive Eschaton: Jürgen Moltmann's Potential Contribution to Animal Theology." *Journal of Animal Ethics* 4, no. 1 (October 24, 2020): 18–36.

McLaughlin, Ryan Patrick. "Jürgen Moltmann (1926–): Creation and Sabbath Theology." In *Animal Theologians*, edited by Andrew Linzey and Clair Linzey, 401–21. NY: Oxford University Press, 2023.

McLaughlin, Ryan Patrick. "Possible Critiques of Cosmocentric Transfiguration." In *Preservation and Protest*, 349–80. Theological Foundations for an Eco-Eschatological Ethics. 1517 Media, 2014. https://doi.org/10.2307/j.ctt9m0sjw.24.

McLaughlin, Ryan Patrick. *Preservation and Protest Theological Foundations for an Eco-Eschatological Ethics*. Emerging Scholars. Minneapolis, Minnesota: Fortress Press, 2014.

McLaughlin, Ryan Patrick. "Redemption in John Wesley's The General Deliverance: Theological Hope vs. Biological Reality." Nampa, ID: Nampa Nazarene University, 2014. https://www.academia.edu/5739806/Redemption_in_John_Wesley_s_The_General_Deliverance_Theological_Hope_vs_Biological_Reality.

Merker, Bjorn. "Consciousness without a Cerebral Cortex: A Challenge for Neuroscience and Medicine." *Behavioral and Brain Sciences* 30, no. 1 (February 2007): 63–81. https://doi.org/10.1017/S0140525X07000891.

Middleton, J. Richard. "Reading Genesis 3 Attentive to Human Evolution: Beyond Concordism and Non-Overlapping Magisteria." In *Evolution and the Fall*, edited by William T. Cavanaugh and James K. A. Smith, 70–92. CHI: Eerdmans, 2017. http://ebookcentral.proquest.com/lib/harvest/detail.action?docID=4865154.

Moltmann, Jürgen. *Creating a Just Future*. Translated by John Bowden. London, UK: SCM Press, 1989.

Moltmann, Jürgen. *Ethics of Hope*. MN: Fortress Press, 2012.

Moltmann, Jürgen. *God in Creation (The Gifford Lectures, 1984–1985)*. MN: Fortress Press, 1993.

Moltmann, Jürgen. *In The End—The Beginning: The Life of Hope*. Translated by Margaret Kohl. MN: Fortress Press, 2004.

Moltmann, Jürgen. *Sun of Righteousness, Arise!: God's Future for Humanity and the Earth*. La Vergne, TN: Hymns Ancient & Modern Ltd, 2010. http://ebookcentral.proquest.com/lib/harvest/detail.action?docID=7430341.

Moltmann, Jürgen. *The Coming of God: Christian Eschatology*. Translated by Margaret Kohl. MN: Fortress Press, 1996.

Moltmann, Jürgen. *The Future of Creation*. Translated by Margaret Kohl. London: SCM Press, 1979.

Moltmann, Jürgen. "The Liberation of the Future and Its Anticipations in History." In *God Will Be In All: The Eschatology of Jürgen Moltmann*, edited by Richard Bauckham, 265–89. MN: Fortress Press, 2001.

Moltmann, Jürgen. *The Source of Life: The Holy Spirit and the Theology of Life*. Translated by Margaret Kohl. London, UK: Fortress Press, 1997.

Moltmann, Jürgen. *The Spirit of Hope: Theology for a World in Peril*. Translated by Margaret Kohl. Louisville, KN: Westminster John Knox Press, 2019.

Moltmann, Jürgen. *The Spirit of Life: A Universal Affirmation*. London: SCM Press, 1992.

Moltmann, Jürgen. *The Way of Jesus Christ: Christology in Messianic Dimensions*. Translated by Margaret Kohl. MN: Fortress Press, 1993.

Muers, Rachel. "The Holy Spirit, the Voices of Nature and Environmental Prophecy." *Scottish Journal of Theology* 67, no. 3 (August 2014): 323–39. http://dx.doi.org.alphacrucis.idm.oclc.org/10.1017/S0036930614000143.

Murata, Kaori, Miho Nagasawa, Tatsushi Onaka, Nobuyuki Kanemaki, Shigeru Nakamura, Kazuo Tsubota, Kazutaka Mogi, and Takefumi Kikusui. "Increase of Tear Volume in Dogs after Reunion with Owners Is Mediated by Oxytocin." *Current Biology* 32, no. 16 (August 22, 2022): R869–70. https://doi.org/10.1016/j.cub.2022.07.031.

Murray, Mary. "The Underdog in History: Serfdom, Slavery and Specis in the Creation and Development of Capitalism." In *Theorizing Animals: Rethinking Humananimal Relations*, edited by Nik Taylor and Tania Signal, 87–106. Leiden, The Netherlands: Brill, 2011.

Murray, Michael J. *Nature Red in Tooth and Claw: Theism and the Problem of Animal Suffering*. Oxford, UK: Oxford University Press, 2008.

Nazianzus, Gregory St. *St Gregory of Nazianzus: On God and Christ: The Five Theological Orations and Two Letters to Cledonius*. Edited by John Behr.

Translated by Lionel Wickham and Frederick Williams. Crestwood, N.Y: St Vladimir's Seminary Press, 2002.

Nel, Marius. "The Notion of the Holy Spirit as Paraclete from a Pentecostal Perspective." *In Die Skriflig* 50, no. 1 (2016): 1–8. http://dx.doi.org.alphacrucis.idm.oclc.org/10.4102/ids.v50i1.2095.

Newton, Jon. "Apocalyptic Ecotheology: A Pentecostal Contribution." In *Pentecostal Missiology and Environmental Degradation*, edited by Eugene Baron and Amos Yong. Carlisle, UK: Langham Global, 2025.

Norris, David S. "Creation Revealed: An Early Pentecostal Hermeneutic." In *The Spirit Renews the Face of the Earth: Pentecostal Forays in Science and Theology of Creation*, edited by Amos Yong, 74–92. Eugene, OR: Pickwick Publications, 2009.

Origen. *On First Principles*. Notre Dame, IN: Ave Maria Press, 2013.

Osborn, Ronald E. *Death before the Fall: Biblical Literalism and the Problem of Animal Suffering*. Downers Grove, IL: IVP Academic, 2014.

Oxford Centre for Animal Ethics. "Andrew Linzey." Accessed April 17, 2021. https://www.oxfordanimalethics.com/who-we-are/director/.

Palmer, Alexandra, Beth Greenhough, Pru Hobson-West, Gail Davies, and Reuben Message. "What Do Scientists Mean When They Talk About Research Animals 'Volunteering'? In: Society & Animals—Ahead of Print." *Society & Animals (Published Online Ahead of Print 2023)*. Accessed August 16, 2023. https://doi-org.alphacrucis.idm.oclc.org/10.1163/15685306-bja10139.

Pannenberg, Wolfhart. *Systematic Theology Vol 2*. Translated by Geoffrey W. Bromiley. London: T&T Clark, 2004.

Pannenberg, Wolfhart, and Geoffrey W. Bromiley. *Systematic Theology Vol 3*. London: T&T Clark, 2004.

Pardo, Michael A., Kurt Fristrup, David S. Lolchuragi, Joyce H. Poole, Petter Granli, Cynthia Moss, Iain Douglas-Hamilton, and George Wittemyer. "African Elephants Address One Another with Individually Specific Name-like Calls." *Nature Ecology and Evolution*, 2024. https://doi.org/10.1038/s41559-024-02420-w.

Payton, James R. *Irenaeus on the Christian Faith: A Condensation of Against Heresies*. Eugene, OR: Wipf and Stock Publishers, 2011. http://ebookcentral.proquest.com/lib/harvest/detail.action?docID=5706733.

Peacocke, Arthur. "The Cost of New Life." In *The Work of Love: Creation as Kenosis*, edited by John C. Polkinghorne. Grand Rapids, MI: Eerdmans, 2001.

Pepperberg, Irene M., Suzanne L. Gray, Shilpa Mody, Francesca M. Cornero, and Susan Carey. "Logical Reasoning by a Grey Parrot? A Case Study of the Disjunctive Syllogism." *Behaviour* 156, no. 5/8 (2019): 409–45.

Perry, David. *Spirit Baptism : The Pentecostal Experience in Theological Focus.* Global Pentecostal and Charismatic Studies. Boston: Brill, 2017. https://search.ebscohost.com/login.aspx?direct=true&db=nlebk&AN=1565681&site=ehost-live.

Poirier, John C. *The Tongues of Angels : The Concept of Angelic Languages in Classical Jewish and Christian Texts.* Wissenschaftliche Untersuchungen Zum Neuen Testament 2. Reihe. Tübingen: Mohr Siebeck GmbH & Co. KG, 2010. https://alphacrucis.idm.oclc.org/login?url=https://search.ebscohost.com/login.aspx?direct=true&db=nlebk&AN=2344366&site=ehost-live.

Pokharel, Sanjeeta, Nachiketha Sharma, and Raman Sukumar. "Viewing the Rare through Public Lenses: Insights into Dead Calf Carrying and Other Thanatological Responses in Asian Elephants Using YouTube Videos." *Royal Society Open Science* 9 (May 18, 2022). https://doi.org/10.1098/rsos.211740.

Polkinghorne, John. *The God of Hope and the End of the World.* New Haven, CT: Yale University Press, 2002. http://ebookcentral.proquest.com/lib/harvest/detail.action?docID=3420227.

Radford Ruether, Rosemary. *Gaia and God: An EcoFeminist Theology of Earth Healing.* NY: Harper One, 1992.

Reed, David A. "Serpent Handling Pentecostal Churches: Revisiting Disputed Borders." Unpublished Paper.

Richie, Tony. "Radical and Responsible." *Journal of Pentecostal Theology* 23 (2014): 216–35. https://doi.org/10.1163/17455251-02301005.

Ritchie, Craig. "Extinction in the Anthropocene: A Critical Analysis." Ph.D., University of Kent (United Kingdom), 2022. https://doi.org/10.22024/UniKent/01.02.99251.

Rizzo, Daniela. "Animal Glossolalia: A Pneumatological Framework for Animal Theology." *Pneuma* 46, no. 1 (April 19, 2024): 60–79. https://doi.org/10.1163/15700747-bja10108.

Rizzo, Daniela. "Beyond Eco-Mission: Towards a Zoological Imperative in Theological Discourse." In Pentecostal Missiology and Environmental Degradation, edited by Eugene Baron and Amos Yong. Langham Global, 2025.

Rizzo, Daniela. "Pentecostal and Evangelical Ecotheologies." In *The Routledge Handbook of Ecotheology*, edited by Matthew Eaton and Timothy Harvie. Routledge, Forthcoming.

Rizzo, Daniela. "The Word Made Flesh and the Spirit's Breath: Bridging Christology and Pneumatology in Animal Theology." *Religions* 15, no. 11 (November 7, 2024): 1355. https://doi.org/10.3390/rel15111355.

Rizzo, Daniela. "Towards a More Holistic Sustainability: Animal Sentience and Spiritual Potential." In *Handbook of Traditional Spirituality and Sustainability*, edited by J. M. Luetz, P Nunn, and W. L. Filho. New York: Springer, 2026.

Robinson, Margaret. "L'nuwey Views of Animal Personhood and Their Implications." In *Animals and Religion*. London: Routledge, 2024.

Rolston, Holmes. "Redeeming a Cruciform Nature." *Zygon: Journal of Religion & Science* 53, no. 3 (September 2018): 739–51. https://doi.org/10.1111/zygo.12428.

Ryan, John P., Danelle E. Cline, John E. Joseph, Tetyana Margolina, Jarrod A. Santora, Raphael M. Kudela, Francisco P. Chavez, et al. "Humpback Whale Song Occurrence Reflects Ecosystem Variability in Feeding and Migratory Habitat of the Northeast Pacific." *PLoS ONE* 14, no. 9 (September 16, 2019): 1–24. https://doi.org/10.1371/journal.pone.0222456.

Sánchez, Leopoldo A. *Receiver, Bearer, and Giver of God's Spirit: Jesus' Life in the Spirit as a Lens for Theology and Life*. Eugene, OR: Wipf and Stock Publishers, 2015. http://ebookcentral.proquest.com/lib/harvest/detail.action?docID=5706660.

Sanders, Cheryl J. "Pentecostal Ethics and the Prosperity Gospel: Is There a Prophet in the House?" In *Afro-Pentecostalism : Black Pentecostal and Charismatic Christianity in History and Culture.*, edited by Estrelda. Y Alexander and Amos Yong. New York: New York University Press, 2011.

Shannon, Laurie. "Invisible Parts: Animals and the Renaissance Anatomies of Human Exceptionalism." In *Animal Encounters*, edited by Tom Tyler and Manuela Rossini, 137–57. Leiden, The Netherlands: Brill, 2009.

Smith, James K A. "Is the Universe Open for Surprise? Pentecostal Ontology and the Spirit of Naturalism." *Zygon: Journal of Religion & Science* 43, no. 4 (December 2008): 879–96.

Smith, James K. A. *Thinking in Tongues: Pentecostal Contributions to Christian Philosophy*. Grand Rapids, MI: Eerdmans, 2010.

Smith, James K. A. "What Stands on the Fall? A Philosophical Exploration." In *Evolution and the Fall*, edited by William T. Cavanaugh and James K. A. Smith, 57–68. Chicago, IL: Eerdmans, 2017.

Sollereder, Bethany N. *God, Evolution and Animal Suffering: Theodicy Without a Fall*. London/ New York: Routledge, 2019.

Southgate, Christopher. "Did the Natural World Go Wrong?" In *God and the Natural World : Theological Explorations in Appreciation of Denis Edwards*, edited by Ted Peters and Marie Turner. Adelaide: ATF Press, 2020.

Southgate, Christopher. *The Groaning of Creation: God, Evolution and the Problem of Evil*. Louisville, Kentucky: Westminster John Knox Press, 2008.

Stone, Jesse D. "Inward Groans and Unknown Tongues: Interpretations of Romans 8.26 in Early Pentecostal Literature." *Journal of Pentecostal Theology* 30, no. 1 (May 5, 2021): 83–102. https://doi.org/10.1163/17455251-30010007.

Stone, Ken. *Reading the Hebrew Bible with Animal Studies*. Stanford, CA: Stanford University Press, 2018.

Swoboda, A. J. "Eco-Glossolalia: Emerging Twenty-First Century Pentecostal and Charismatic Ecotheology." *Rural Theology* 9, no. 2 (2011): 101–16. https://doi.org/10.1558/ruth.v9i2.101.

Swoboda, A. J. "Greening Hollenweger: Retrieving Ecological Dimensions from the Dean of Pentecostal Studies." In *Pentecostal Missiology and Environmental Degradation*, edited by Eugene Baron and Amos Yong. Carlisle, UK: Langham Global, 2025.

Swoboda, A. J. "Looking the Wrong Way: Salvation and the Spirit in Pentecostal Eco-Theology." In *A Liberating Spirit: Pentecostals and Social Action in North America*, edited by Michael Wilkinson and Steven M Studebaker, 231–47. Eugene, OR: Pickwick Publications, 2010.

Swoboda, A. J. "Reconciling Creation: Spirit, Salvation, and Ecological Degradation." *Australasian Pentecostal Studies* 22, no. 1 (June 14, 2021): 87–103.

Swoboda, A. J. *Tongues and Trees– Towards a Pentecostal Ecological Theology*. Dorset: Deo Publishing, 2013.

Torr, Stephen C. "Lamenting in Tongues: Glossolalia as a Pneumatic Aid to Lament," March 17, 2017. https://doi.org/10.1163/17455251-02601005.

Van Ruler, Arnold A. *Calvinist Trinitarianism and Theocentric Politics: Essays Toward a Public Theology*. Translated by John Bolt. Vol. 38. Toronto Studies in Theology. Lewiston, NY: The Edwin Mellen Press, 1989.

Vondey, Wolfgang. *Beyond Pentecostalism: The Crisis of Global Christianity and the Renewal of the Theological Agenda*. Grand Rapids, MI: Eerdmans, 2010.

Vondey, Wolfgang. "Pentecostal Theology." *Journal of Pentecostal Theology* 28 (2019): 32–45. https://doi.org/10.1163/17455251-02801004.

Waldrop, Richard E. "Spirit of Creation, Spirit of Pentecost: Reflections on Ecotheology and Mission in Latin American Pentecostalism." In *Blood Cries Out: Pentecostals, Ecology, and the Groans of Creation*, edited by A. J. Swoboda, 225–34. Eugene, Oregon: Pickwick Publications, 2014.

Wallace, Mark I. *Finding God in the Singing River: Christianity, Spirit, Nature*. MI: Fortress Press, 2005.

Wallace, Mark I. *Fragments of the Spirit: Nature, Violence and the Renewal of Creation*. New York: Continuum Publishing Company, 1996.

Wallace, Mark I. *When God Was a Bird: Christianity, Animism, and the Re-Enchantment of the World*. New York: Fordham University Press, 2019.

Walsh, Froma. "Human-Animal Bonds I: The Relational Significance of Companion Animals." *Family Process* 48, no. 4 (December 2009): 462–80. https://doi.org/10.1111/j.1545-5300.2009.01296.x.

Wang, Connie X., Isaac A. Hilburn, Daw-An Wu, Yuki Mizuhara, Christopher P. Cousté, Jacob N. H. Abrahams, Sam E. Bernstein, Ayumu Matani, Shinsuke Shimojo, and Joseph L. Kirschvink. "Transduction of the Geomagnetic Field as Evidenced from Alpha-Band Activity in the Human Brain." *eNeuro* 6, no. 2 (March 1, 2019). https://doi.org/10.1523/ENEURO.0483-18.2019.

Wariboko, Nimi. *The Pentecostal Principle: Ethical Methodology in New Spirit*. Grand Rapids, MI: Eerdmans, 2012.

Way, Kenneth C. "Animals in the Prophetic World: Literary Reflections on Numbers 22 and 1 Kings 13." *Journal for the Study of the Old Testament* 34, no. 1 (September 2009): 47–62. https://doi.org/10.1177/0309089209346353.

Webb, Stephen H. *Dome of Eden: A New Solution to the Problem of Creation and Evolution*. Eugene, OR: Cascade Books, 2010.

Webb, Stephen H. *On God and Dogs: A Christian Theology of Compassion for Animals*. New York: Oxford University Press, 2002.

Welker, Michael. *God the Spirit*. Translated by John Hoffmeyer. Minneapolis: Fortress Press, 1992.

Wenk, Matthias. "An Incarnational Pneumatology Based on Romans 8:18–30: The Spirit as God's Solidarity with a Suffering Creation." *Religions* 13, no. 3 (2022): 191. https://doi.org/10.3390/rel13030191.

Wesley, John. "Sermon 60: The General Deliverance." In *The Works of John Wesley*, 3rd ed., 5:241–52. Grand Rapids, MI: Baker Book House, 1991.

Woodley, Randy S. *Indigenous Theology and the Western Worldview: A Decolonized Approach to Christian Doctrine*. Grand Rapids, MI: Baker Academic, 2022.

Yong, Amos. *In the Days of Caesar: Pentecostalism and Political Theology*. Grand Rapids, MI: Eerdmans, 2010.

Yong, Amos. *Renewing Christian Theology: Systematics for a Global Christianity*. Waco, Texas: Baylor University Press, 2014.

Yong, Amos. *Spirit-Word-Community: Theological Hermeneutics in Trinitarian Perspective*. Eugene, OR: Wipf & Stock Publishers, 2002.

Yong, Amos. *The Cosmic Breath: Spirit and Nature in the Christianity-Buddhism-Science Trialogue*. Leiden; Boston: Brill, 2012.

Yong, Amos. "The Missio Spiritus: Towards a Pneumatological Missiology of Creation." In *Creation Care in Christian Mission*, edited by Koama Kapya. Regnum Edinburgh Centenary Series. Oxford : Regnum Books International, 2015.

Yong, Amos. *The Spirit of Creation: Modern Science and Divine Action in the Pentecostal—Charismatic Imagination*. Grand Rapids, MI: Eerdmans, 2011.

Yong, Amos. *The Spirit Poured Out on All Flesh: Pentecostalism and the Possibility of Global Theology*. MI: Baker Academic, 2005.

Yong, Amos, ed. *The Spirit Renews the Face of the Earth : Pentecostal Forays in Science and Theology of Creation*. Oregon: Pickwick Publications, 2009.

Yong, Amos. "The Truth of Tongues Speech: A Rejoinder to Frank Macchia." *Journal of Pentecostal Theology* 6, no. 13 (October 1998): 106–15. https://doi.org/10.1177/096673699800601306.

Yong, Amos. "Translation as Social and Ecological Justice (A Review Essay): Notes toward a Theology of Tongues and Interpretations in the Age of the Anthropocene." *Pneuma*, no. Forthcoming (2025).

York, Tripp. "Can the Wolf Lie Down with the Lamb without Killing It? Confronting the Not-So-Practical Politics of the Peaceable Kingdom." In *A Faith Embracing All Creatures*, edited by Tripp York and Andy Alexis-Baker, 150–65. Eugene, OR: Cascade Books, 2012.

York, Tripp, and Laura Hobgood-Oster. *The End of Captivity?: A Primate's Reflections on Zoos, Conservation, and Christian Ethics*. Eugene, Oregon: Cascade Books, 2015.

Index

D. Rizzo, *Animal Pneumatology*, The Palgrave Macmillan Animal Ethics Series, https://doi.org/10.1007/978-3-032-18723-9

The manufacturer's authorised representative in the EU is Springer Nature Customer Service Centre GmbH, Europaplatz 3, 69115 Heidelberg, Germany. If you have any concerns regarding our products, please contact ProductSafety@springernature.com

Printed and bound by CPI Group (UK) Ltd, Croydon, CR0 4YY
21/09/2026
02219962-0001